American Book Company
THE STANDARDS EXPERTS

North Carolina Online Testing is now available!

On your computer: 4 Easy Steps!

1. Visit americanbookcompany.com/takeatest
2. Select your state and grade
3. Select NC READY Biology
4. Enter the password "genetics"

On your mobile device: 4 Easy Steps!

1. Scan the QR code below
2. Select your state and grade
3. Select NC READY Biology
4. Enter the password "genetics"

Biology

NC Ready

2014-2015 Edition

Scan this QR code with your smart device to jump to the online testing page.

Online testing available through: August 1, 2015

North Carolina READY End-of-Course Assessment in Biology

2014-2015 Edition

Michelle Gunter

American Book Company
PO Box 2638
Woodstock, GA 30188-1383
Toll Free: 1 (888) 264-5877 Phone: (770) 928-2834
Fax: (770) 928-7483 Toll Free Fax: 1 (866) 827-3240
Website: www.americanbookcompany.com

ACKNOWLEDGEMENTS

The author would like to gratefully acknowledge the formatting and technical expertise of Becky Wright and Marsha Torrens as well as the editing and proofreading contributions of Susan Barrows.

I also want to thank Mary Stoddard and Eric Field for their expertise in developing the graphics for this book.

Thank you to Phil Lausier for allowing us the use of his photographs.

PO Box 2638
Woodstock, GA 30188-1318

Printed in the United States of America

04/14

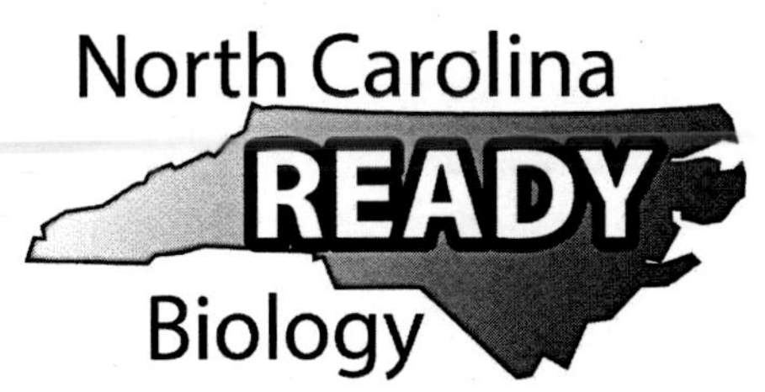

Table of Contents

Preface

North Carolina READY End-of-Course Assessment in Biology will help students who are learning or reviewing material for the READY standards in biology. The materials in this book are based on the READY standards as published by the *North Carolina Department of Public Instruction*.

This book contains several sections. These sections are as follows: 1) General information about the book; 2) A Pretest; 3) An Evaluation Chart; 4) Chapters that teach the concepts and skills that improve graduation readiness; 5) A Post Test. Answers to the tests and exercises are in a separate manual. The answer manual also contains a Chart of Standards for teachers to make a more precise diagnosis of student needs and assignments.

We welcome comments and suggestions about the book. Please contact the author at

American Book Company
PO Box 2638
Woodstock, GA 30188-1383

Toll Free: 1 (888) 264-5877
Phone: (770) 928-2834
Fax: (770) 928-7483
Web site: www.americanbookcompany.com

About the Author

Michelle Gunter graduated from Kennesaw State University in Kennesaw, Georgia, with a B.S. in Secondary Biology Education. She is a certified teacher in the field of biology in the state of Georgia. She has three years' experience in high school science classrooms. She has nine years' experience in biology and biological systems. She has won awards for her research in the field of aquatic toxicology. Mrs. Gunter enjoys teaching students of all ages the wonders of the natural world.

TEST-TAKING TIPS

1 **Complete the chapters and practice tests in this book.** This text will help you review the skills for the North Carolina READY End-of-Course Assessment in Biology.

2 **Be prepared.** Get a good night's sleep the day before your exam. Eat a well-balanced meal, one that contains plenty of proteins and carbohydrates, prior to your exam.

3 **Arrive early.** Allow yourself at least 15–20 minutes to find your room and get settled. Then you can relax before the exam so you won't feel rushed.

4 **Think success.** Keep your thoughts positive. Turn negative thoughts into positive ones. Tell yourself you will do well on the exam.

5 **Practice relaxation techniques.** Some students become overly worried about exams. Before or during the test, they may perspire heavily, experience an upset stomach or have shortness of breath. If you feel any of these symptoms, talk to a close friend or see a counselor. They will suggest ways to deal with test anxiety. **Here are some quick ways to relieve test anxiety:**

- Imagine yourself in your most favorite place. Let yourself sit there and relax.
- Do a body scan. Tense and relax each part of your body starting with your toes and ending with your forehead.
- Use the 3 – 12 – 6 method of relaxation when you feel stress. Inhale slowly for 3 seconds. Hold your breath for 12 seconds, and then exhale slowly for 6 seconds.

6 **Read directions carefully.** If you don't understand them, ask the proctor for further explanation before the exam starts.

7 **Use your best approach for answering the questions.** Some test-takers like to skim the questions and answers before reading the problem or passage. Others prefer to work the problem or read the passage before looking at the answers. Decide which approach works best for you.

8 **Answer each question on the exam.** Unless you are instructed not to, make sure you answer every question. If you are not sure of an answer, take an educated guess. Eliminate choices that are definitely wrong, and then choose from the remaining answers.

9 **Use your answer sheet correctly.** Make sure the number on your question matches the number on your answer sheet. In this way, you will record your answers correctly. If you need to change your answer, erase it completely. Smudges or stray marks may affect the grading of your exams, particularly if they are scored by a computer. If your answers are on a computerized grading sheet, make sure the answers are dark. The computerized scanner may skip over answers that are too light.

10 **Check your answers.** Review your exam to make sure you have chosen the best responses. Change answers only if you are sure they are wrong.

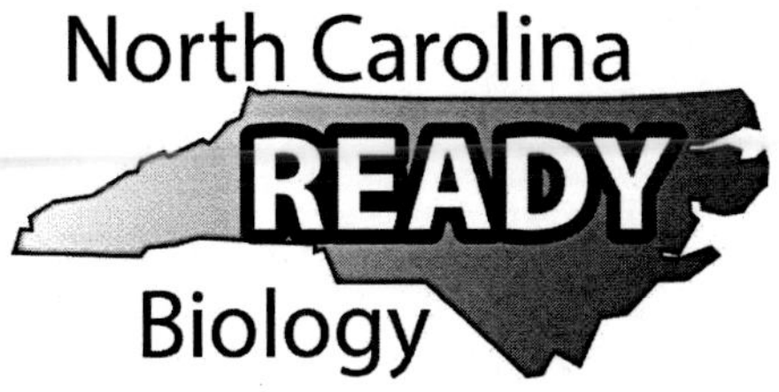

Chart of Standards

<table>
<tr><th>Standard:
Chapter</th><th>Pretest</th><th>Post Test</th></tr>
<tr><td colspan="3">1.1.1 Summarize the structure and function of organelles in eukaryotic cells (including: the nucleus, plasma membrane, cell wall, mitochondria, vacuoles, chloroplasts, and ribosomes) and ways that these organelles interact with each other to perform the function of the cell.</td></tr>
<tr><td>1</td><td>2, 7, 8, 12, 20, 33</td><td>48</td></tr>
<tr><td colspan="3">1.1.2 Compare prokaryotic and eukaryotic cells in terms of their general structure (plasma membrane and genetic material) and degree of complexity.</td></tr>
<tr><td>1</td><td>—</td><td>9, 53, 56</td></tr>
<tr><td colspan="3">1.1.3 Explain how instructions in DNA lead to cell differentiation and result in cells specialized to perform specific function in multicellular organisms.</td></tr>
<tr><td>1</td><td>18, 37, 54</td><td>1, 2, 42, 59</td></tr>
<tr><td colspan="3">1.2.1 Explain how homeostasis is maintained in the cell and within an organism in various environments (including: temperature and pH)</td></tr>
<tr><td>2</td><td>24, 25, 27</td><td>—</td></tr>
<tr><td colspan="3">1.2.2 Analyze how cells grow and reproduce in terms of interphase, mitosis and cytokinesis.</td></tr>
<tr><td>2</td><td>—</td><td>12, 15, 55</td></tr>
<tr><td colspan="3">1.2.3 Explain how specific cell adaptations help cells survive in particular environments (focus on unicellular organisms).</td></tr>
<tr><td>2</td><td>21, 51</td><td>50</td></tr>
</table>

Standard: Chapter	Pretest	Post Test
2.1.1 Compare the flow of energy and cycling of matter (water, carbon, nitrogen and oxygen) through ecosystems relating the significance of each to maintaining the health and sustainability of an ecosystem.		
3	26, 44	3, 6
2.1.2 Analyze the survival and reproductive success of organisms in terms of behavioral, structural, and reproductive adaptations.		
3	49, 50	30, 37, 46, 47
2.1.3 Explain various ways organisms interact with each other (including predation, competition, parasitism, mutualism) and with their environments resulting in stability within ecosystems.		
3	39	10, 14, 25, 40
2.1.4 Explain why ecosystems can be relatively stable over hundreds or thousands of years, even though populations may fluctuate (emphasizing availability of food, availability of shelter, number of predators and disease).		
3	5, 17, 46	—
2.2.1 Infer how human activities (including population growth, pollution, global warming, burning of fossil fuels, habitat destruction and introduction of nonnative species) may impact the environment.		
4	15	33
2.2.2 Explain how the use, protection and conservation of natural resources by humans impact the environment from one generation to the next.		
4	3, 22	54

Standard: Chapter	Pretest	Post Test
3.1.1 Explain the double-stranded, complementary nature of DNA as related to its function in the cell.		
5	28	4, 57
3.1.2 Explain how DNA and RNA code for protein and determine traits.		
5	13	7, 22, 43, 60
3.1.3 Explain how mutations in DNA that result from interactions with the environment (i.e. radiation, and chemicals) or new combinations in existing genes lead to changes in function and phenotype.		
5	53, 57, 60	13
3.2.1 Explain the role of meiosis in sexual reproduction and genetic variation.		
6	19	23, 45
3.2.2 Predict offspring ratios based on a variety of inheritance patterns (including: dominance, co-dominance, incomplete dominance, multiple alleles, and sex-linked traits).		
6	4, 11, 14, 36	20, 21, 24
3.2.3 Explain how the environment can influence the expression of genetic traits.		
6	18, 38	58
3.3.1 Interpret how DNA is used for comparison and identification of organisms.		
7	6, 16	5, 16, 52
3.3.2 Summarize how transgenic organisms are engineered to benefit society.		
7	47	8, 26
3.3.3 Evaluate some of the ethical issues surrounding the use of DNA technology (including: cloning, genetically modified organisms, stem cell research, and Human Genome Project).		
7	—	49
3.4.1 Explain how fossils, biochemical, and anatomic evidence support the theory of evolution.		
8	31, 32, 40, 41, 42, 43, 48, 56	29
3.4.2 Explain how natural selection influences the changes in species over time.		
8	9	17, 31, 34, 36
3.4.3 Explain how various disease agents (bacteria, viruses, chemicals) can influence natural selection.		
8	—	—
3.5.1 Explain the historical development and changing nature of classification systems.		
9	23	11, 38
3.5.2 Analyze the classification of organisms according to their evolutionary relationships (including: dichotomous keys and phylogenetic trees).		
9	10, 34	35

Standard: Chapter	Pretest	Post Test
4.1.1 Compare the structures and functions of the major biological molecules (carbohydrates, proteins, lipids, and nucleic acids) as related to the survival of living organisms.		
10	29, 35, 52	18
4.1.2 Summarize the relationship among DNA, proteins, and amino acids in carrying out the work of cells and how this is similar in all organisms.		
10	55, 59	19, 32
4.1.3 Explain how enzymes act as catalysts for biological reactions.		
10	—	28
4.2.1 Analyze photosynthesis and cellular respiration in terms of how energy is stored, released, and transferred within and between these systems.		
11	30, 45, 58	27, 39, 41, 44, 51
4.2.2 Summarize ways that organisms use released energy for maintaining homeostasis (active transport).		
11	1	—

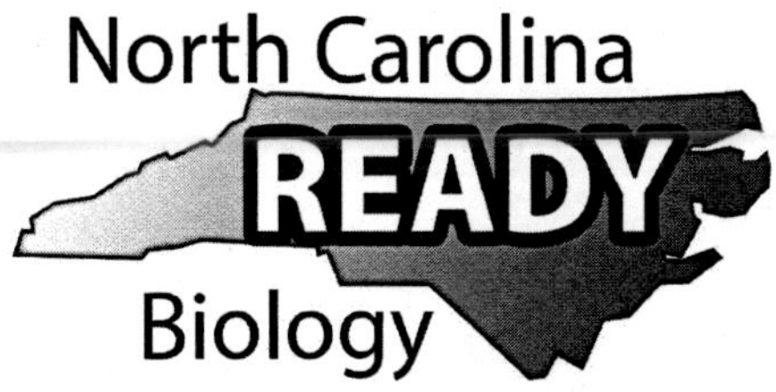

Pretest

1 Which of the following illustrates active transport? 4.2.2

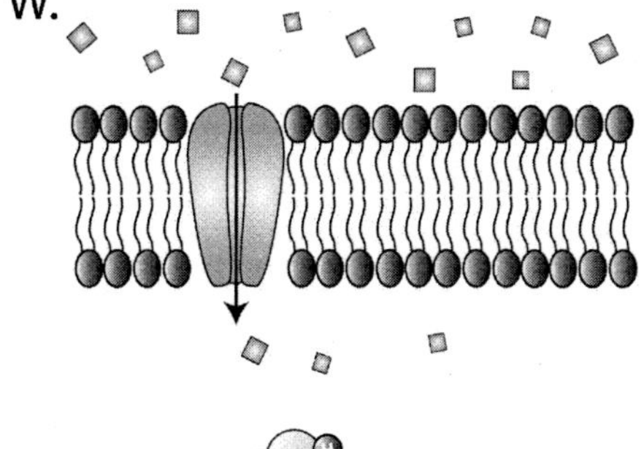

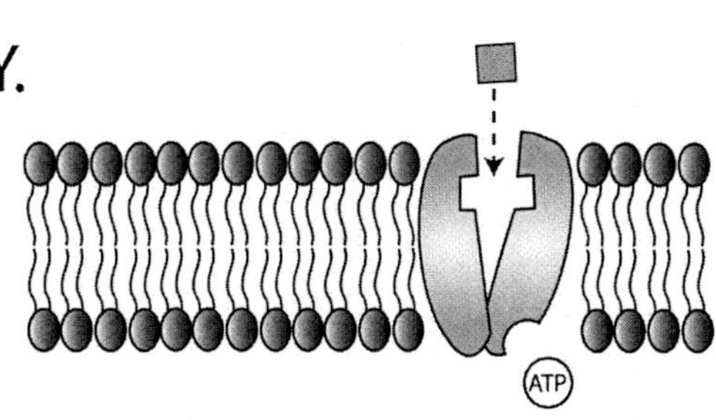

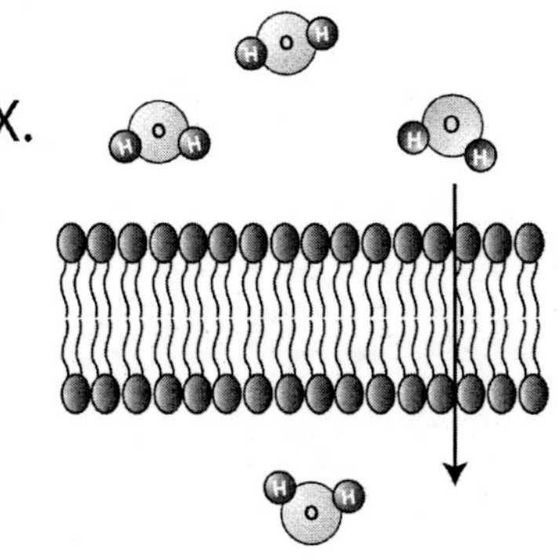

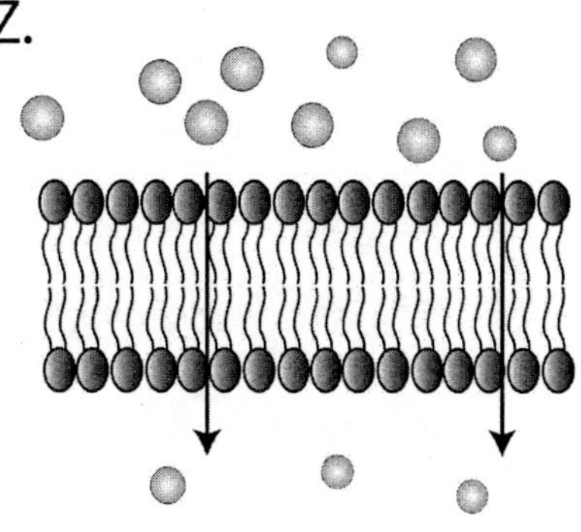

A Image W

B Image X

C Image Y

D Image Z

2 Manfred was interested in cells. He wondered how a cell's function affected the number of organelles each cell contained. He knows that muscle cells use lots of energy. Manfred looked at different types of body cells under a microscope. Look at his data below. 1.1.1

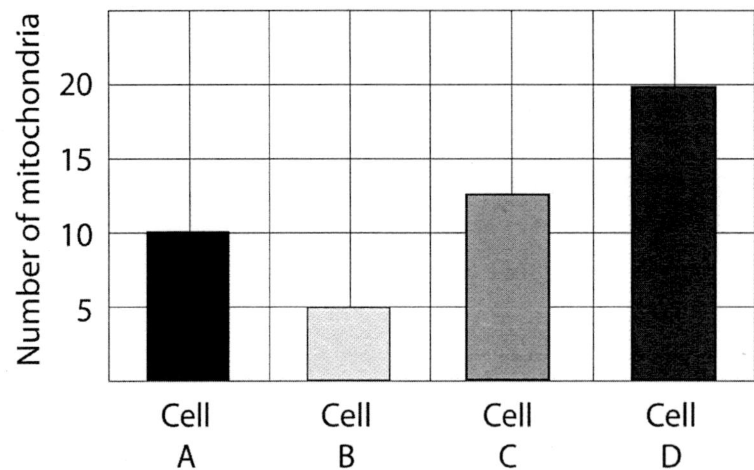

What cell do you predict ***most likely*** came from a muscle?

A Cell A

B Cell B

C Cell C

D Cell D

3 The use of windmills to produce power is encouraged by many environmental proponents and by the US government, which subsidizes this power technology at a comparatively high rate. Windmills do not pollute; they produce a great deal of clean energy and their use does not deplete any natural resource. Wind power is a good example of technology that harnesses which kind of resource? 2.2.2

A a nonrenewable resource

B a renewable resource

C a fossil fuel

D a biofuel

Go to next page

4 Red and blue are two popular variations of a particular flower. A gardener began experimental breeding with this flower. His results are shown below: 3.2.2

The First Generation of Flowers:

Color	Total Percent of Offspring
Red	0%
Blue	0%
Purple	100%

The Second Generation of Flowers:

Color	Total Percent of Offspring
Red	25%
Blue	25%
Purple	50%

What inheritance pattern is seen in these flowers?

A simple dominant/recessive

B incomplete dominance

C dominance

D communicable

5 In a grassland ecosystem, sunlight is abundant but moisture is limited. In a forest ecosystem just the opposite is true, sunlight is limited but moisture is abundant. How do you think these facts impact the ecological succession from grassland to forest ecosystem? 2.1.4

A Plants that survive in a grassland system can also survive in a forest system.

B Plants that survive in a grassland system cannot survive in a forest system.

C Animals cannot survive under these changing conditions.

D Animals can survive in all systems.

6 During a murder trial, criminal forensic prosecutors used DNA technology to match seed pods found in the suspect's pickup truck to one specific tree growing near the victim's unmarked grave. Which technique did prosecutors ***most likely*** use to conclusively link the suspect to the victim? 3.3.1

A transgenic organisms

B bacterial transformation

C gel electrophoresis

D cloning

7 Proteins are made in the ribosomes of a cell, whereas DNA is found only in 1.1.1

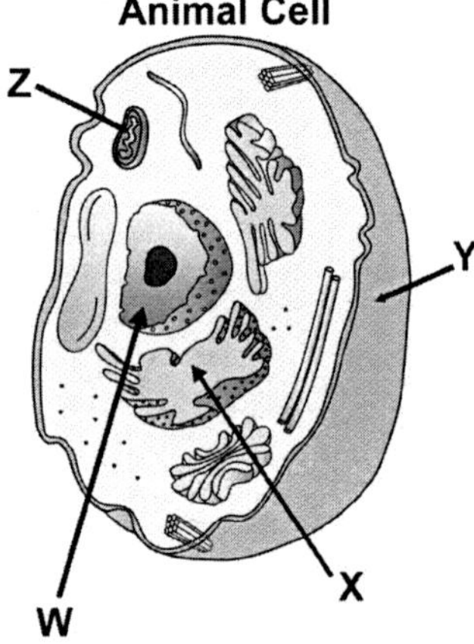

A organelle W.

B organelle X.

C organelle Y.

D organelle Z.

8 Which energy-producing organelle is indicated by the arrow in the diagram below? 1.1.1

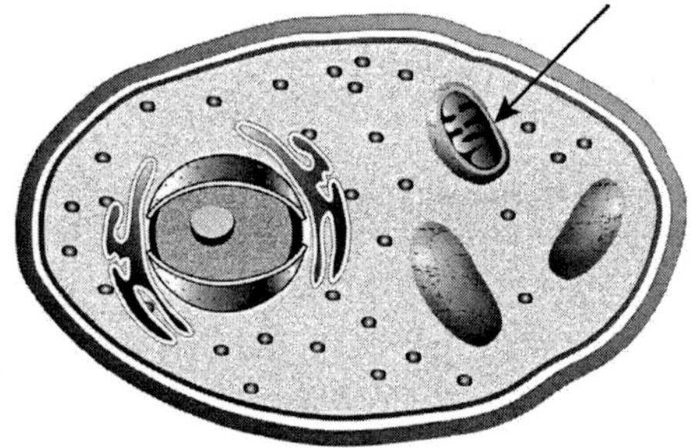

A ribosomes

B nucleolus

C nucleus

D mitochondria

9 An adult female American alligator (*Alligator mississippiensis)* lays 50 eggs each breeding season. Of those offspring, less than 1% will survive to adulthood. This information supports what part of Darwin's evolutionary theory? 3.4.2

A Only the strong survive.

B Organisms produce more offspring than the environment can support.

C Genetic mutation is a common route for acquiring inherited characteristics.

D Characteristics acquired by an individual during its lifetime are often passed on to offspring.

10 According to the classification key, to which genus and species does this cat belong? 3.5.2

1. a. Fur is marked with stripes and spots ....go to 2.
   b. Fur is primarily solid-colored ....go to 3.
2. a. Dwells primarily in trees ....*Leopardus wiedii* (margay)
   b. Dwells primarily on the ground .... *Leopardus pardalis* (ocelot)
3. a. Large animal with black-tipped ears and tail ....*Puma concolor* (puma)
   b. Medium animal with long, black, tufted ears .... *Caracal caracal* (caracal)

A *Leopardus wiedii*

B *Leopardus pardalis*

C *Puma concolor*

D *Caracal caracal*

Go to next page

11 A chicken has two kinds of genes that determine the color of its plumage: one lends the chicken a background color and the other modifies that background. Two common plumage genes are E and I. E is a gene coding for background color; it is often called "extended black." I is a modifying gene called "dominant white." The I gene is codominant to E. What color will a chicken with the gene combination EI be? 3.2.2

A It will be gray.

B It will be completely black.

C It will be completely white.

D It will be black with white spots.

12 A cell has large vacuoles, chloroplasts and a cell wall. What kind of cell is this ***most likely*** to be? 1.1.1

A a plant cell

B an animal cell

C neither a plant or an animal cell

D a prokaryotic cell

13 Which situation results in an inheritable characteristic? 3.1.2

A A turtle eats an arm off a starfish.

B A horse learns to unlock a gate with its mouth.

C A jellyfish gene is transferred into rice seed allowing rice to produce more vitamin A.

D A random mutation causes the immediate death of an embryo.

14 In mice, brown hair is dominant to white. Cross a heterozygous female with a heterozygous male. The phenotypic ratio will yield 3.2.2

A 3 brown: 1 white.

B 3 white: 1 brown.

C 1 brown: 2 tan: 1 white.

D 100% brown hair.

Go to next page

15 English ivy is an evergreen vine. It grows on top of any surface. It grows up tree trunks and over branches toward sunlight. Ramiro hypothesized that if English ivy grew on top of a native producer, then the native producer would die. Look at his data below. 2.2.1

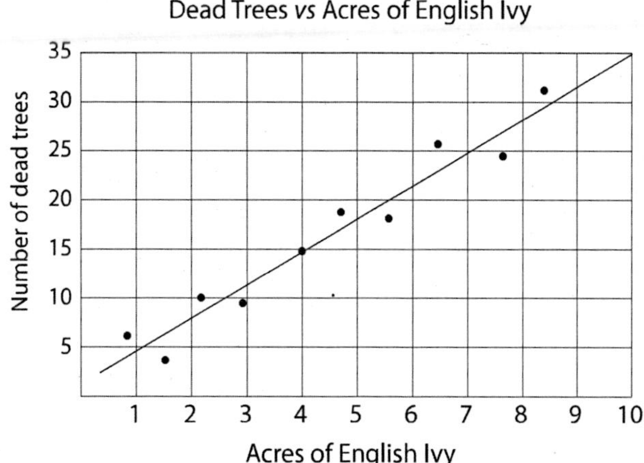

Which statement ***best*** describes what he should conclude about the native producers in his hypothesis?

A His hypothesis was supported; English ivy reduced the native producers in an area.

B His hypothesis was unsupported; English ivy reduced the native producers in an area.

C His hypothesis was unsupported; English ivy increased the native producers in an area.

D His hypothesis was proved correct; English ivy increased the native producers in an area.

Go to next page

16 A liquor store is burglarized at twilight. After the police arrive, they find blood on broken glass at the scene. A DNA analysis of three suspects produces the genetic fingerprints shown below. Which suspect was the burglar? 3.3.1

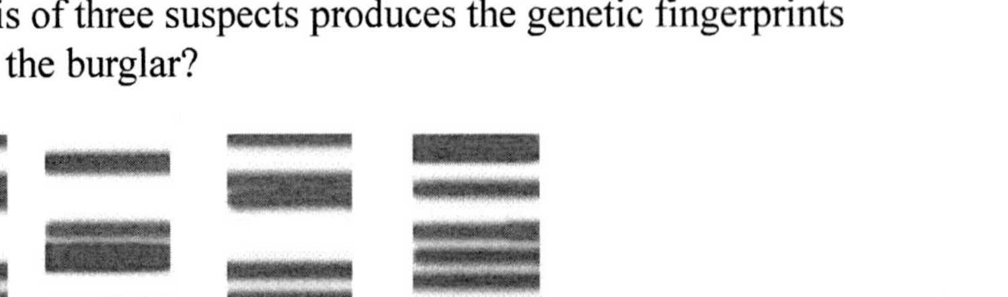

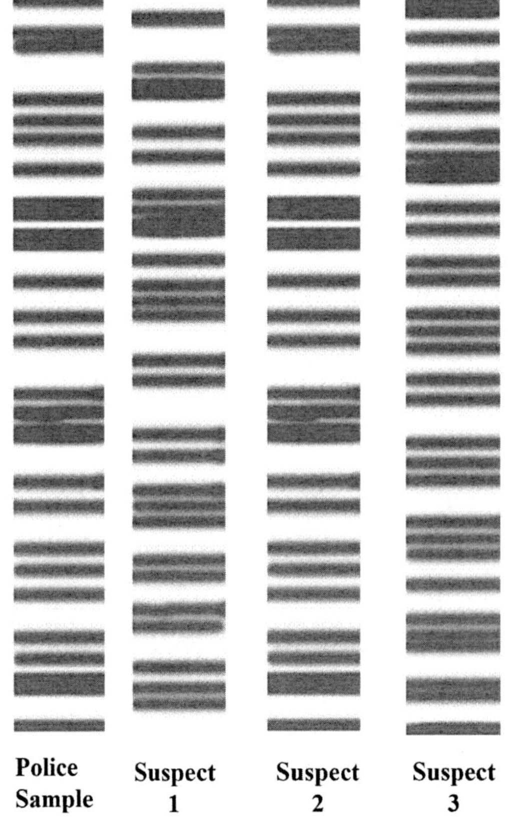

A Suspect 1

B Suspect 2

C Suspect 3

D All suspects have some of the DNA markers from the police sample, so the test is inconclusive.

Go to next page

17 A new emerging disease kills native birds. It is a density-dependent limiting factor for bird populations. Students conducted population estimates for four bird species in their area. 2.1.4

Species	# of Individuals per Acre
A	12
B	68
C	7
D	19

Which species will be ***most*** impacted by this new disease?

A Species A

B Species B

C Species C

D Species D

18 During the development of a new embryo the temperature rises above 120 °C, well above tolerable limits. How is this environmental change ***most likely*** to affect cell differentiation? 3.2.3, 1.1.3

A Cell differentiation will be positively affected in some way.

B Cell differentiation will likely speed up.

C Cell differentiation will not be affected and continue normally.

D Cell differentiation will be negatively affected in some way.

19 Which process of cell division results in progeny with half the chromosomes of the parent cells? 3.2.1

A meiosis

B replication

C mitosis

D fertilization

20 The organelle indicated in the diagram contains a pigment responsible for capturing sunlight needed for which process? 1.1.1

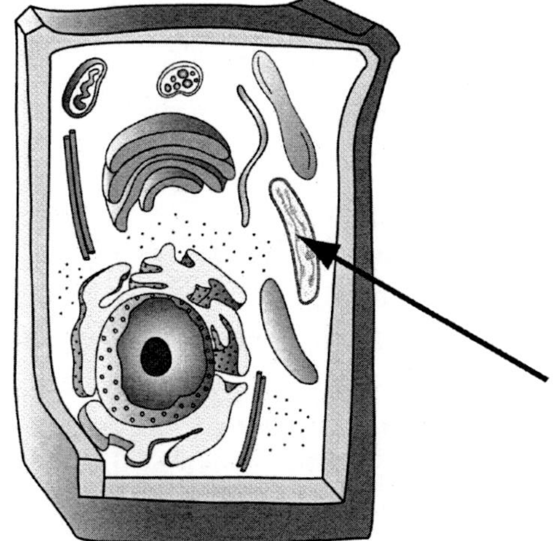

A photosynthesis

B aerobic respiration

C nutrient absorption

D cellular transport

21 Which adaptation listed below ***best*** helps this freshwater organism to survive in a salty environment? 1.2.3

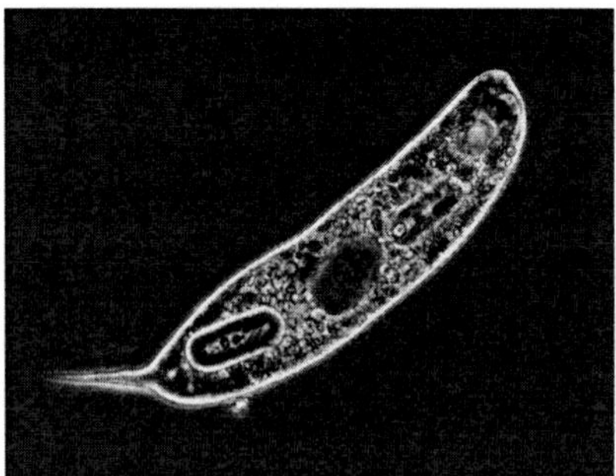

A nucleus

B endoplastic reticulum

C ribosome

D contractile vacuole

22 The African savanna has a large range of highly specialized plants and animals, which depend on each other to keep the environment in balance. In many parts of the savannas, the African people have begun to graze their livestock. What is the ***likely*** outcome of this activity? 2.2.2

A The savanna grasses will grow more quickly as they are eaten, so the area of the savanna will increase.

B The top consumers will leave the area, as there are no more animals to eat.

C The grasses will be diminished and the entire ecosystem will be affected.

D The loss of vegetation will cause ground-water to overflow, so the savanna biome will convert to a flooded grassland.

23 The diagram shows the most current classification of birds of prey. It takes into account recent genetic evidence that suggests these birds all arose from a common ancestor. 3.5.1

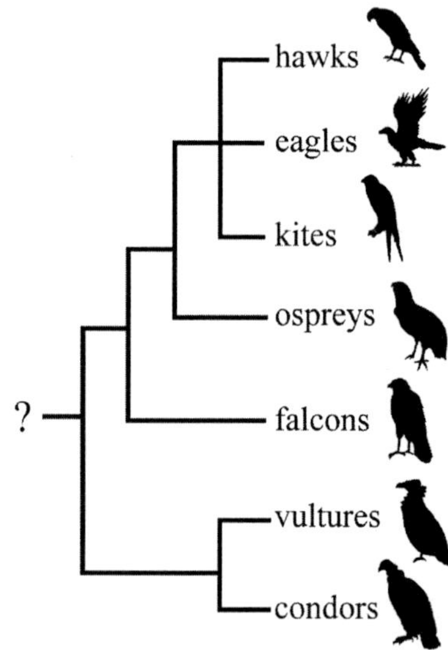

Which statement below offers the ***best*** explanation for how different birds of prey arose from a single ancestral group?

A The offspring of weaker birds were physically unable to pass along necessary genetic traits.

B Environmental variations selected for genetic change within the population.

C Stronger birds hunted faster and better than weaker birds in harsh habitats.

D Larger birds had more dominant traits than recessive birds.

Go to next page

24 Examine the experimental setup below. 1.2.1

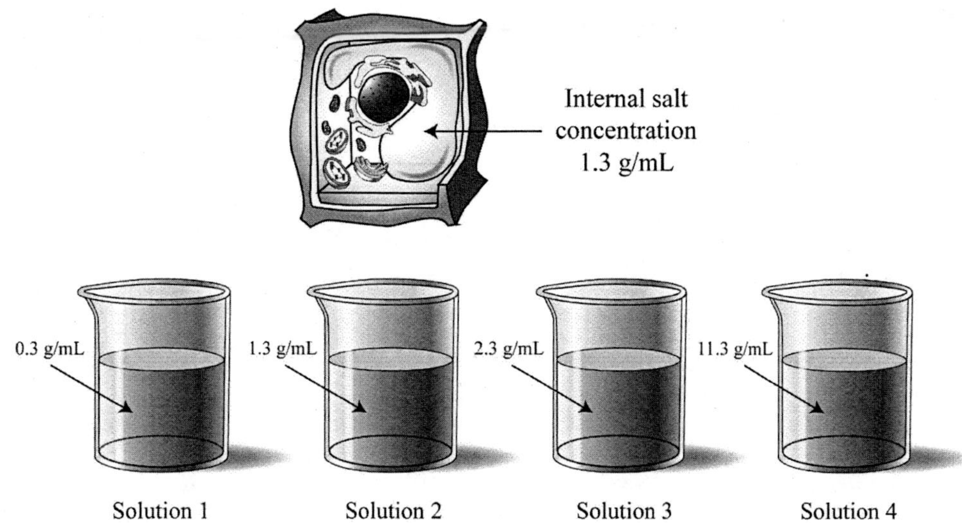

Which solution will ***most*** increase the osmotic pressure experienced by the cell?

A Solution 1

B Solution 2

C Solution 3

D Solution 4

Go to next page

25 A freshwater plant is moved to a salt marsh. Predict the effect of that movement on the plant. 1.2.1

A The plant will wilt.

B The plant will turn red.

C The plant will swell.

D The plant will not change its water concentration.

26 Which of the following statements concerning physical and biological systems is correct? 2.1.1

A Matter recycles in a given system, while energy flows through it.

B Energy recycles in a given system, while matter flows through it.

C Matter recycles only in physical systems, not in biological systems.

D Energy flows only through biological systems, not physical systems.

27 An animal cell is placed in a hypertonic solution. What will ***most likely*** happen to the cell? 1.2.1

A It will shrink.

B It will swell.

C It will shrink then swell.

D It will remain the same size.

28 If all 30,000 human genes are made up of the same 20 nucleotides, how does one gene differ from another? 3.1.1

A the size of the nitrogen bases

B the type of bonds between the nucleotides

C the sequence of the nitrogen bases

D the phenotypes of the nitrogen bases

29 One class of biological molecules consists of chains of methyl (–CH) units that are long or short. What are these called? 4.1.1

A carbohydrates

B lipids

C enzymes

D proteins

30 In what way are aerobic and anaerobic respiration most similar? 4.2.1

A They both require a high temperature environment.

B They both happen inside all types of cells.

C They both function with oxygen.

D They both produce ATP.

Go to next page

31 The table below shows some traits of four organisms. 3.4.1

Animal	Jaw bone	Native Range	Number of Legs
Cheetah	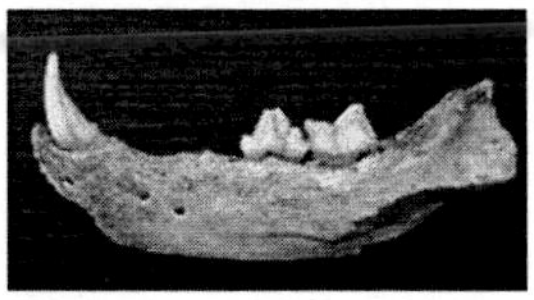	Africa	4
Mountain Lion	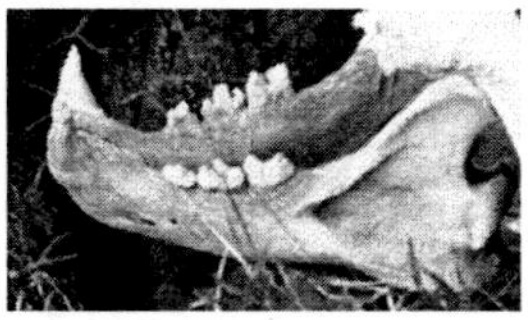	North America	4
Horse	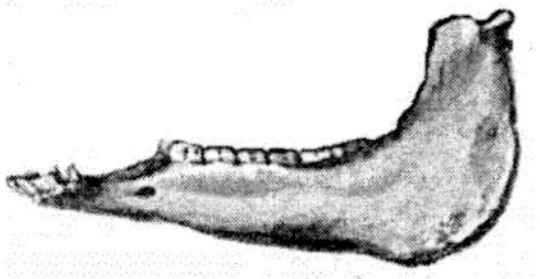	Europe	4
Rattlesnake	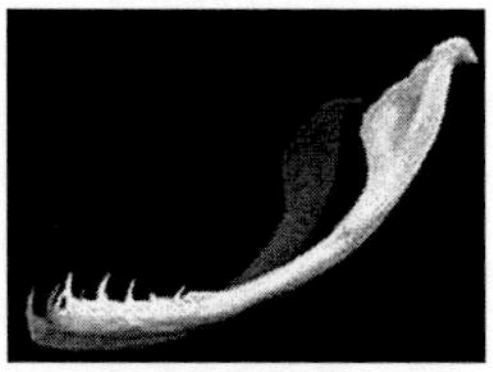	North America	0

Which two organisms are ***most*** related?

A mountain lion and horse

B rattlesnake and cheetah

C horse and rattlesnake

D cheetah and mountain lion

32 Study these four paired graphics and identify which two structures are ***most*** analogous. 3.4.1

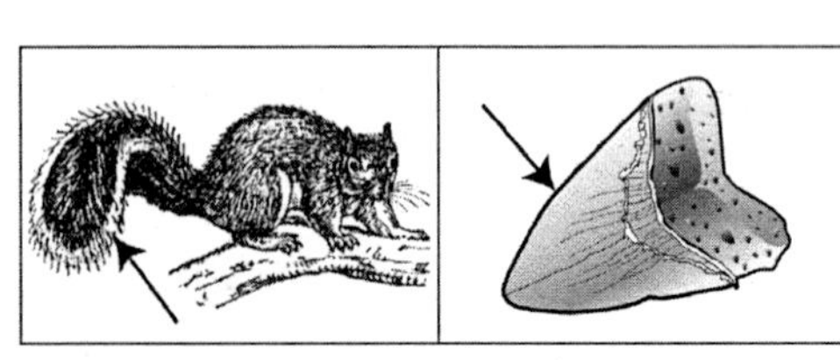

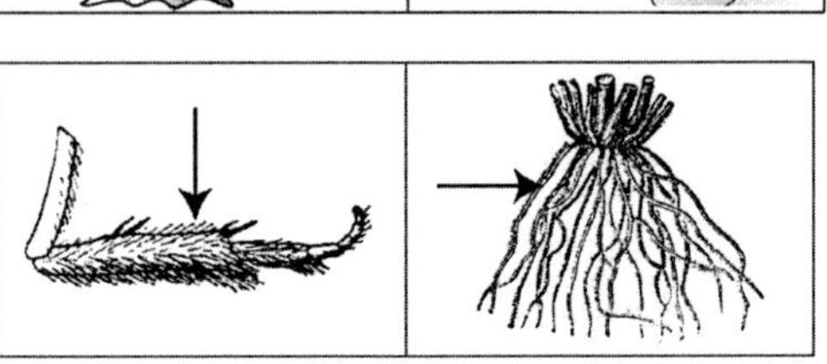

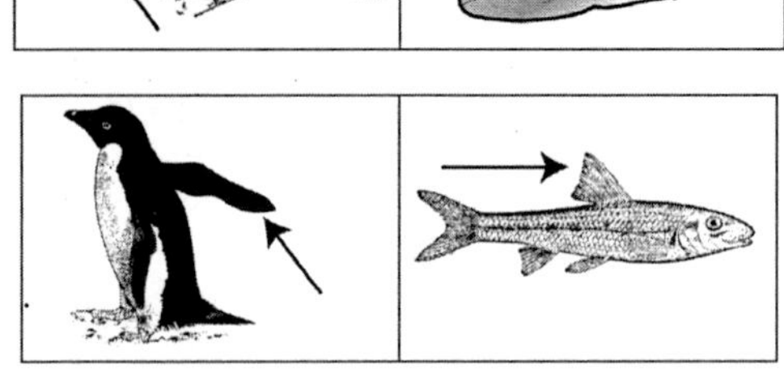

A tree trunk and bird wing

B insect leg and plant root

C squirrel tail and shark tooth

D penguin wing and fish fin

Go to next page

33 This diagram represents a cell. 1.1.1

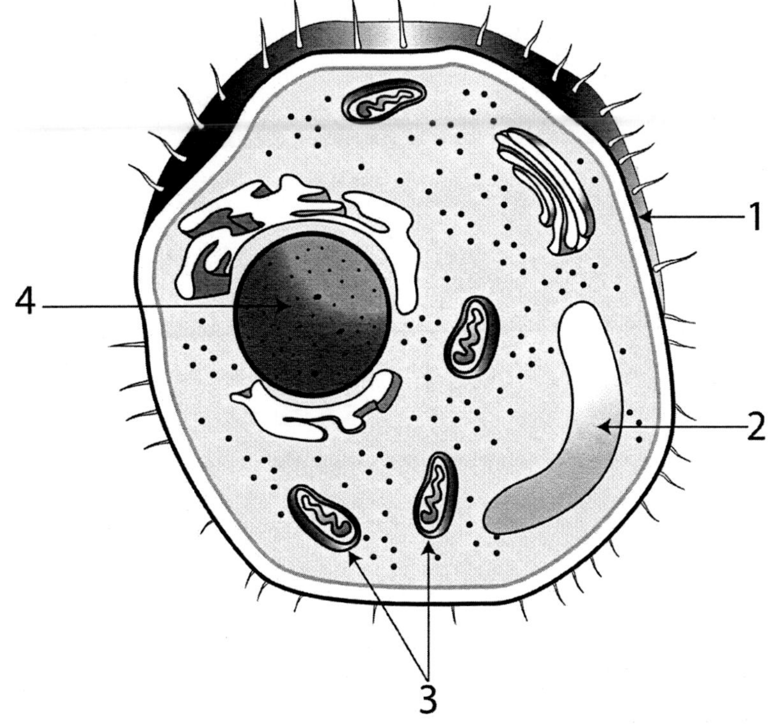

Which cellular part is the site where ADP is converted into ATP?

A Structure 1

B Structure 2

C Structure 3

D Structure 4

34 Which two organisms are ***most closely*** related? 3.5.2

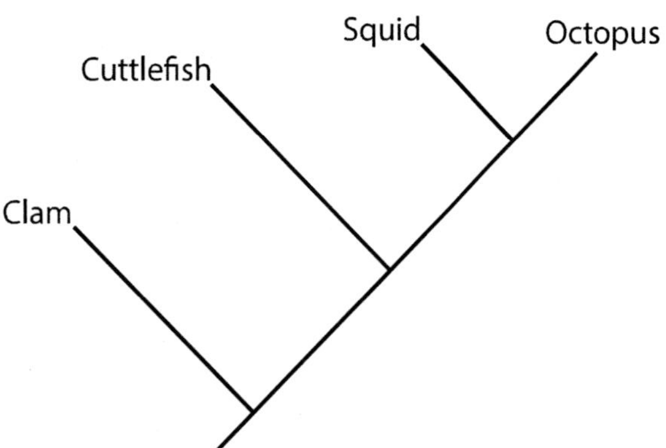

A clam and octopus

B cuttlefish and octopus

C clam and squid

D squid and octopus

35 What is the main function of carbohydrates within a cell? 4.1.1

A build the cell membrane

B provide cellular energy

C store cellular information

D provide water storage within the cell

36 A landscaper crossed two heterozygous junipers. She noticed that of the offspring junipers, 73% were short and 27% were tall. What do these results indicate about the allele for shortness? 3.2.2

A dominant

B recessive

C codominant

D incompletely dominant

37 Puppies often ger roundworms in their intestines. the worms consume some of the food the puppies have eaten. The worms and the puppies are in a relationship know as 1.1.3

A mutualism

B competition

C predation

D parasitism

Go to next page

38 The peppered moth comes in two speckled colorations: a recessive light coloration and a dominant dark coloration. The graphs below show changes to the moth population during a 50 year period. 3.2.3

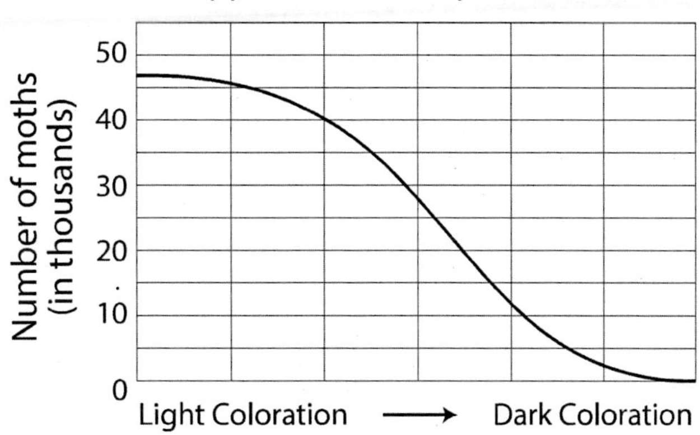

Peppered Moth Population 1900

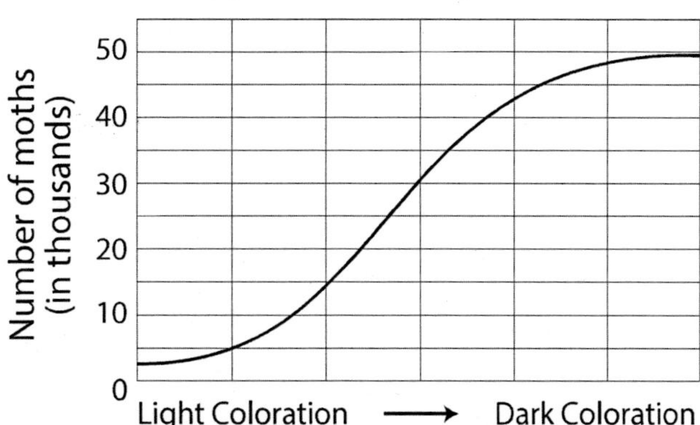

Which of the following statements most ***accurately*** represents the trend demonstrated by the peppered moths?

A A smaller percentage of light-colored moths survive to reproduce, shrinking the gene pool and causing mutations.

B More and more dark-colored moths survive to reproduce, which shifted the allele frequency towards the dark-colored allele.

C Over time, the light-colored moths will become homozygous for the light allele and become extinct.

D Over time, the birds will get used to eating the light-colored moths and stop eating the dark-colored moths.

Go to next page

39 Mandy observed some organisms in a container. She noticed Organism 2 growing only on top of Organism 1. Look at her data below. 2.1.3

	Day 1	**Day 5**	**Day 10**	**Day 15**
Organism 1	large green foliage	yellow foliage	brown foliage	died
Organism 2	small white fibers	more small white fibers growing on organism #1	lots of small white fibers on organism #1	fewer small white fibers

Which statement ***best*** describes the relationship between these two organisms?

A Both species benefit and neither is harmed.

B Only one species benefits and the other remains unharmed.

C One species is harmed while the other benefits.

D Both species change to suit the new environment.

40 The skulls shown are from four different animals. Which two animals are the ***most closely*** related? 3.4.1

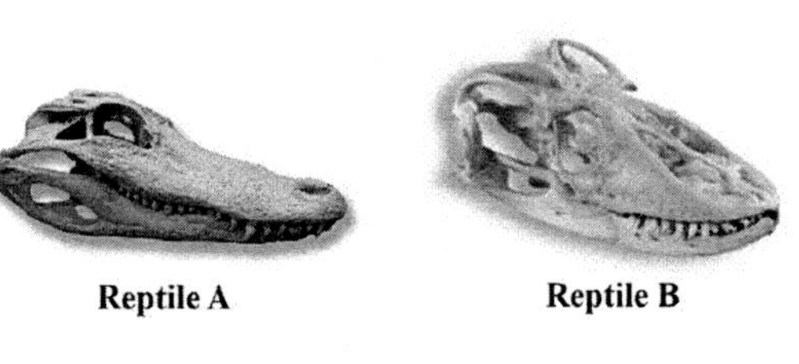

Reptile A Reptile B

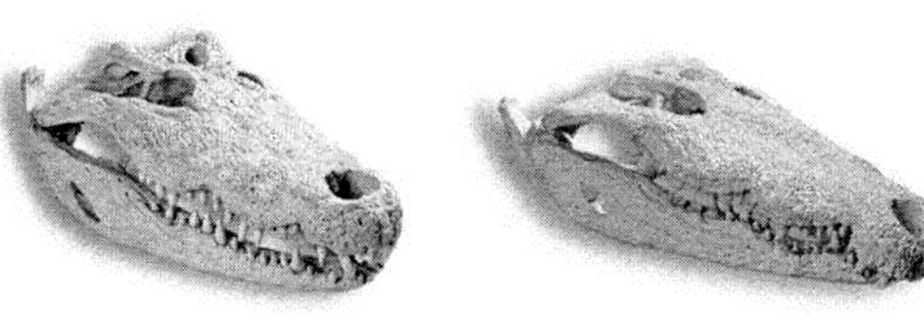

Reptile C Reptile D

A Reptiles A and B

B Reptiles B and C

C Reptiles A and D

D Reptiles C and D

41 A unit of heredity, composed of a segment of DNA on a chromosome that contains information required to manufacture a protein and will eventually be expressed as as trait, is called a 3.4.1

A ribosome.

B chromatin.

C base.

D gene.

42 Predict the cross of a homozygous tall parent witha homozygous short parent if tall is dominant over short. The resultes of the offspring will be 3.4.1

A all tall

B all short

C neither short nor tall

D some tall and some short.

43 Mutations occurring in the reproductive cells 3.4.1

A affect only the tissues of the organism affected.

B may be passed on to future descendents.

C are never beneficial.

D rarely result in genetic disorders.

44 Which two groups of organisms listed below are ***most*** responsible for recycling decomposing organic matter within the ecosystem? 2.1.1

A plants and animals

B animals and invertebrates

C bacteria and fungi

D animals and fungi

45 During strenuous exercise, animals cannot take in enough air to supply their cells with the necessary oxygen. They begin to carry out lactic acid fermentation to supply their cells with the necessary energy. Lactic acid fermentation is a type of 4.2.1

A anaerobic respiration.

B aerobic respiration.

C alcoholic respiration.

D ATP respiration.

46 How will rapid environmental change impact the stability of an ecosystem? 2.1.4

A It will make the ecosystem more stable.

B It will make the ecosystem less stable.

C It will not impact the stability of the ecosystem.

D It will help the ecosystem attract more animals.

47 Transgenic plants have DNA from another organism inserted into its genome. What is one main benefit of using transgenic plants in crop production? 3.3.2

A Transgenic plants may cause allergic reactions in people.

B Transgenic plants may reduce world hunger and malnutrition.

C Transgenic plants may create unknown wild hybrids.

D Transgenic plants may have unintended consequences in ecosystems.

Go to next page

48 An experiment is conducted to discover the effect of a protein (F1) being inactivated in chicks prior to hatching. The procedure developed involves administering an inhibiting virus to a chick during its incubation. Four different species of bird chicks (A – D) are used. The results are shown below. 3.4.1

	Webbed feet?	Feathers on feet?
Experimental Bird A	yes	no
Normal Bird A	yes	no
Experimental Bird B	yes	yes
Normal Bird B	no	no
Experimental Bird C	no	yes
Normal Bird C	no	yes
Experimental Bird D	yes	yes
Normal Bird D	no	no

What can you conclude about the relatedness of these four bird species, based on this data?

A Birds A and D are likely related.

B Birds C and D are likely related.

C Birds B and D are likely related.

D Birds A and B are likely related.

Go to next page

49 Consider the plants shown. Based on their root structures, which plant is better adapted to life in a desert environment? 2.1.2

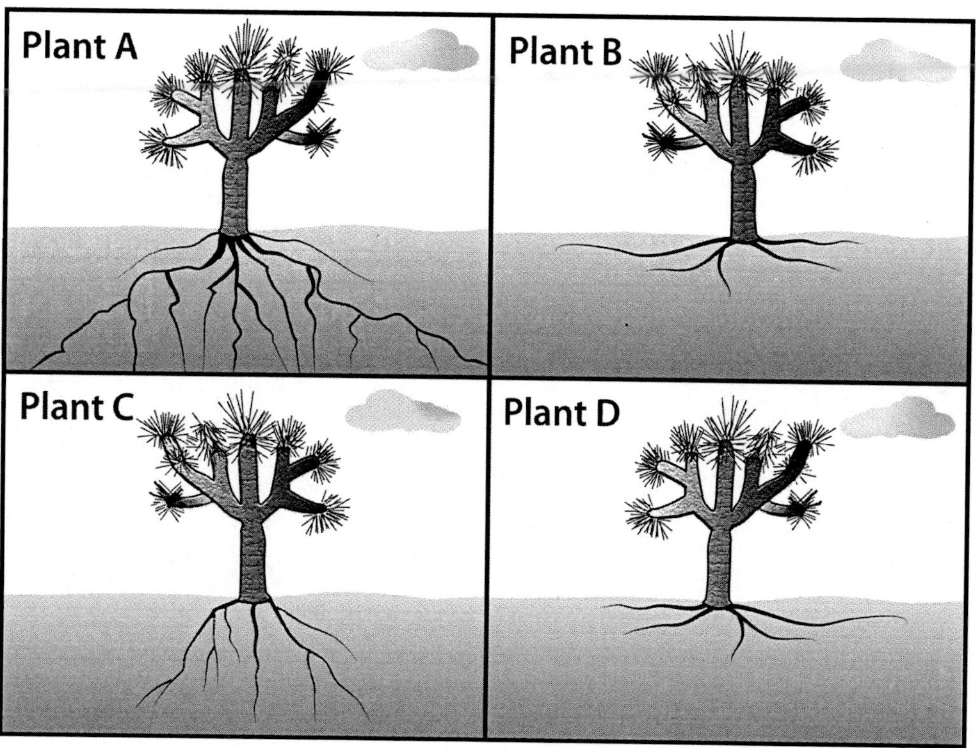

A Plant A

B Plant B

C Plant C

D Plant D

50 Which physical feature below makes sea turtles more adapted to life in the ocean than life on land? 2.1.2

A large protective shell

B flipperlike legs

C amniotic egg

D short tail

51 The euglena has a structure called an eyespot. This structure senses light. Which statement ***best*** explains how this structure helps the euglena survive? 1.2.3

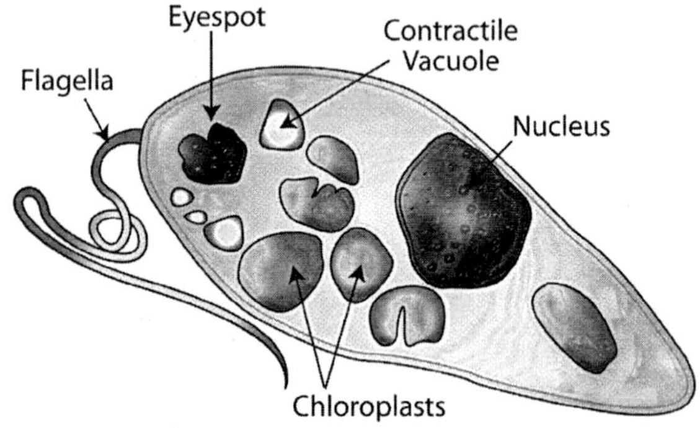

A It helps keep the euglena near sunlight to perform photosynthesis.

B It helps keep the euglena away from sunlight to perform respiration.

C It helps the euglena keep the correct amount of salt in its cytoplasm.

D It provides the euglena with a means to escape from predators.

Go to next page

52 Which image below ***best*** represents a phospholipid? 4.1.1

A

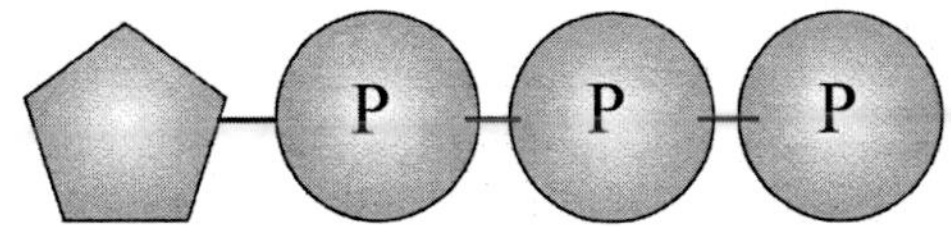

C

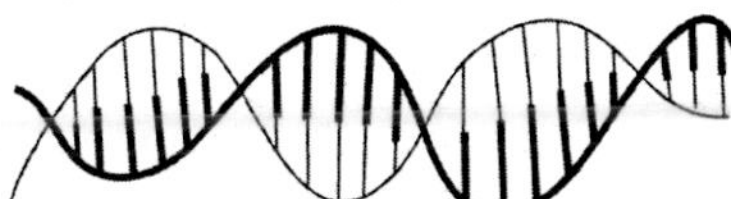

B

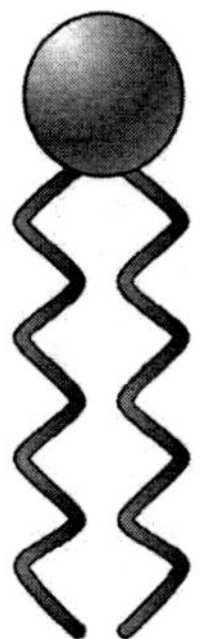

D

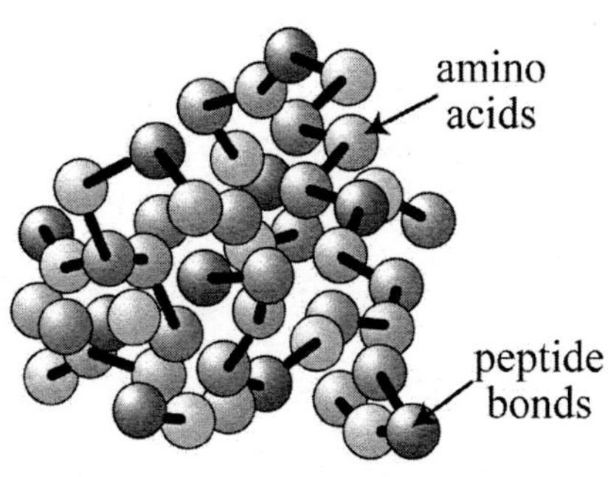

Go to next page

53 Which factor below is ***most likely*** to cause a mutation in the DNA of a cell? 3.1.3

A exposure to temperatures between 40 °F and 75 °F

B exposure to liquid H_2O

C exposure to radiation

D exposure to dust

54 An animal cell that is specialized to create movement is ***most likely*** found in which organ system? 1.1.3

A the nervous system

B the circulatory system

C the skeletal system

D the muscular system

55 What is the last step of protein synthesis? 4.1.2

A amino acids linking together

B tRNA moving to the ribosome

C rewriting DNA into RNA

D mRNA moving out of the nucleus

56 Which comparison would give a scientist the ***most*** information about the relatedness of two species? 3.4.1

A comparing their DNA

B comparing their adaptation

C comparing their body covering

D comparing their growth rate

57 Military-grade bioweapons are microbial biological diseases that have been "heated up" through mutations in their DNA. Bioweapons are more virulent forms of natural diseases. Which factor below do you predict was ***most likely*** used to create these types of bioweapons? 3.1.3

A radiation

B temperature extremes

C salt solutions

D hormones

58 In what way are photosynthesis and respiration similar? 4.2.1

A They use the same reactants and have the same products.

B They involve energy transformation.

C They happen inside bacterial cells.

D They require light to proceed.

59 Which substance listed below is one of the five nitrogenous bases? 4.1.2

A cytosol

B adenosine

C thymine

D ribose

Go to next page

60 A particular gene that codes for coat pigment is extremely temperature sensitive. Warm or hot temperatures cause the gene to switch off. Cooler temperatures allow the gene to be expressed, resulting in a dark coloration of the animal's coat. If a cat had this particular gene for coat color, select the ***best*** representation of its coloration. 3.1.3

A Cat A

B Cat B

C Cat C

D Cat D

North Carolina READY Biology Evaluation Chart

Chapter	Questions
Scientific Inquiry Review	–
Chapter 1: Cellular Structure	2, 7, 8, 12, 18, 20, 33, 37, 54
Chapter 2: Cellular Function	21, 24, 25, 27, 51
Chapter 3: Ecosystem	5, 17, 26, 39, 44, 46, 49, 50
Chapter 4: Human Actions	3, 15, 22
Chapter 5: DNA	13, 28, 53, 57, 60
Chapter 6: Genetic	4, 11, 14, 18, 19, 36, 38
Chapter 7: Biotechnology	6, 16, 47
Chapter 8: Evolution	9, 31, 32, 40, 41, 42, 43, 48, 56
Chapter 9: Classification	10, 23, 34
Chapter 10: Biomolecule	29, 35, 52, 55, 59
Chapter 11: Energy	1, 30, 45, 58

Scientific Inquiry Review *

*This section addresses aspects of scientific inquiry not directly covered in the NC Biology Standards.

SCIENTIFIC PROCESS

Every day, you have an opportunity to think scientifically. For instance, on a warm summer night, you may step outside and notice all the different life forms buzzing around you. You might notice lightning bugs flashing in the dark or moths swarming around a street lamp. You may say to yourself, "Gee, these bugs sure are pretty," and go about your business. On the other hand, you might start asking questions about the insects.

Kinds of Insects

- Why do lightning bugs glow?
- Why are moths attracted to light?
- What makes some bugs active only at night?

These are **questions** that deal directly with your **observation**. You probably could find the answers to these kinds of questions in an encyclopedia. You might even ask questions that require some effort on your part to answer.

- How do lightning bugs make light?
- How do moths find food in the dark?
- What advantages are there to being active at night?

In even simple, everyday observations, there are literally hundreds of questions that can be asked. Thinking scientifically also means pursuing the answers to these questions.

A functional definition of science is the observation, identification, description and explanation of **natural phenomena**. Natural phenomena are observable facts or events in the world around us, like the insects. Scientific processes help explain natural phenomena. In fact, for a question to be considered a scientific question, you must be able to answer it in a scientific way through observation, testing and analysis. Scientists believe all natural phenomena have logical, verifiable explanations — sometimes it just takes some thought, effort and time to find them!

Scientific Inquiry Review *

Through the study of science, we ask questions, develop **hypotheses** (educated guesses) and design and carry out experiments to gain a better understanding of the universe. Then we must try to make sense of the experimental results through analysis. Only then can we arrive at some conclusion about our hypothesis.

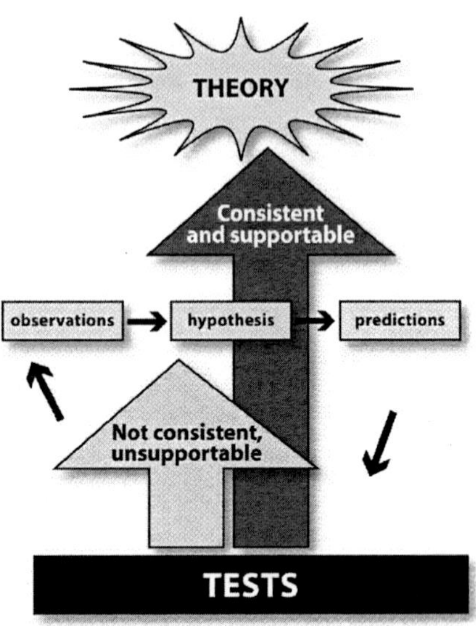

The System of Scientific Thought

To "do" science, you must have some way of thinking through the possible solutions to a problem and testing each possibility to find the best solution. There are many valid scientific processes, and an entire book could be written about how thought is translated into questions, experiments and conclusions. But since our space for this topic is limited, we'll provide you with one of the most common "scientific processes." Our scientific process will have the following steps:

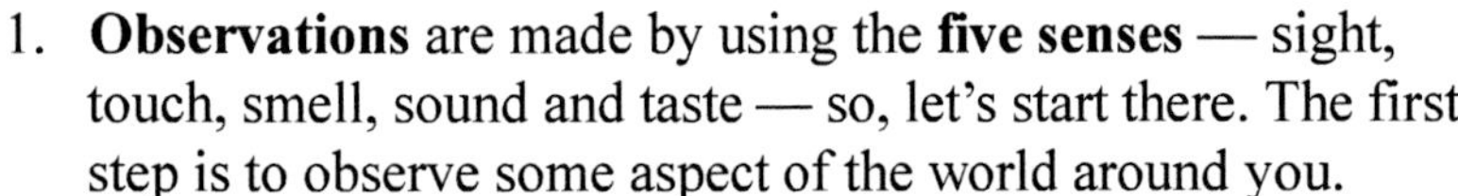

1. **Observations** are made by using the **five senses** — sight, touch, smell, sound and taste — so, let's start there. The first step is to observe some aspect of the world around you.
2. **Identify a problem or question** to solve based on your observations. How can you state the problem as a question for investigation?
3. Do a little **research** to find out what is already known about your question or topic.
   - use only reliable sources
4. State a **hypothesis** — that is another way of saying "an educated guess about the possible solution to your problem."
   - inductive reasoning allows you to draw on personal experience
   - deductive reasoning requires you to use a well-accepted truth or theory
   - can be in the form of "If... then..." statement
5. Conduct an **experiment** or set of experiments that aims to produce results that will support or contradict your hypothesis.
   - independent variables are the variables that you change during the experiment
   - dependent variables are the variables or factors that you measure
   - control variables are all other variables in the experiment that must be held constant
6. Collect and organize your **data**. What does it tell you?
   - qualitative data assesses the qualities of your observations
   - quantitative data allows you to collect numerical figures
   - use tables, line graphs, pie charts and bar graphs to simplify data
7. **Analyze** the data and summarize the results as a conclusion in terms of the original hypothesis.
   - can be used to generate inferences, predictions and models

Remember, this is just one way to organize your thoughts; things don't always happen this way. Sometimes you might have a hypothesis and only later on realize a problem for which it is applicable. Your experiment may generate unexpected data, which is hard to interpret and doesn't answer your question as clearly as you were hoping. What do you do then? There are several paths you could take. You could redo your experiment or, perhaps rework the experimental setup. Lastly, you might decide to conduct a new experiment based on the new information.

Section Review 1: Scientific Process

A. Define the following terms.

observation	natural phenomena	hypothesis	five senses
dependent variable	inductive reasoning	chart	conclusion
control variable	quantitative data	scientific experiment	inference
experimental group	diagram	reproducible	model
control group	table	variable	prediction
qualitative data	graph	independent variable	

B. Choose the best answer.

1. Identify the statement below that can be answered in a scientific way.

   A Apples are more delicious than pears.

   B Snails are beautiful invertebrates.

   C Diamonds are the most valuable substance on earth.

   D Romaine lettuce is more nutritious than iceberg lettuce.

2. What is a hypothesis?

   A an educated guess

   B a natural phenomena

   C something seen or heard

   D a scientific experiment

3. How are theories developed?

   A by making observations with the five senses

   B by thinking of one scientific hypothesis

   C through scientific predictions

   D through many consistent and supportable tests

C. Complete the following exercise.

1. Make some scientific observations, and write at least one hypothesis.

EQUIPMENT AND MEASUREMENT

Laboratory equipment and materials are tools used in scientific investigations. Each piece of equipment in the lab has a specific purpose. Part of behaving safely is the ability to identify lab equipment and recognize its use. Inappropriate use of equipment and material causes accidents. If you don't know the purpose of a piece of equipment, ask your teacher. Instruments are used to enhance the ability of our senses to observe.

	Types of Equipment	SI Unit	Base Unit
Measuring Mass	analytical pan balance, triple-beam balance	milligram (mg), gram (g), kilogram (kg)	kilogram (kg)
Measuring Length	meter stick, metric ruler, caliper	millimeter (mm), meter (m), kilometer (km)	meter (m)
Measuring Volume	graduated cylinder, pipet	cubic centimeter (cm^3), milliliter (mL) liter (L)	liter (L)
Observation Equipment	microscope, telescope, magnifying glass	nanometers (nm), kilometer (km), micrometers (μm), light year, millimeters (mm)	
Heating Equipment	Bunsen burner, hot plate, tongs, tripod, wire gauze	Celsius (°C), Fahrenheit (°F), Kelvin (K)	Kelvin (K)
Handling Liquids	Erlenmeyer flask, test tubes, beakers, eyedropper	cubic centimeter (cm^3), milliliter (mL) liter (L)	

SAFETY IN THE LABORATORY

Safety procedures are set up to protect you and others from injury. Hopefully, you will never become injured in a science lab! The most important safety rule is to *always follow your teacher's instructions.* Before working in the laboratory, fully read all of the directions for the experiment. Laboratory accidents can be easily avoided if safety procedures are followed. Be sure to dress appropriately for the laboratory environment. Know where eyewash stations and safety showers are located. Determine what personal protective equipment, like aprons, goggles and/or gloves, are necessary to ensure the safety of yourself and others. If there is an accident, spill or breakage in the laboratory, report it to your instructor immediately. *REMEMBER: You must be concerned about your own safety as well as the safety of others working around you.*

Manufacturers of chemicals are required to produce, update and maintain a safety data sheet for each chemical they produce. This document is called a **material safety data sheet**, or **MSDS**. An MSDS lists information on chemical structure, chemical appearance, chemical properties and personal safety. It also contains information on safe storage and disposal of chemicals.

An MSDS comes with every hazardous chemical that is purchased. In fact, there is an MSDS for every chemical known, even water and air! MSDS records are not discarded, but kept on file. If you want to see the MSDS for a chemical that you are working with, ask your teacher.

QUICK REVIEW

1. Large amounts of petrified wood are found in northeast Arizona. Using inductive reasoning, four inferences are made. Which is the *most* reasonable?

   A All wood becomes petrified.

   B A living forest once stood there.

   C No forests grew in other parts of Arizona.

   D Wood only becomes petrified in northeast Arizona.

2. Which of the following is ***most likely*** a reliable source for scientific information?

   A *National Geographic*

   B *New England Journal of Medicine*

   C *Scientific American*

   D *Ladies' Home Journal*

3. If you were to do an experiment, which of the following would be the ***best*** hypothesis to test?

   A Do goldfish like warm water or cold water?

   B Goldfish are more active in warm water than in cold water.

   C Goldfish live in warm and cold water.

   D Temperature changes will kill goldfish.

4. Which of the following phrases contains quantitative data?

   A Green leaves surround white flowers.

   B Ricky's football jersey is number 85.

   C Seeds sprout more quickly when it is warm.

   D Water evaporated at a rate of 2 mL per minute.

5. Identify a piece of equipment used to measure mass.

   A spring scale

   B triple-beam balance

   C caliper

   D graduated cylinder

6. Kilograms are a unit of measurement for

   A mass. B height. C volume. D size.

Scientific Inquiry Review *

7. Which science tool would be ***best*** to use to measure the length of a caterpillar?

   A ruler

   B beaker

   C scale

   D volume

8. What is the standard SI unit used to measure mass?

   A kilograms

   B grams

   C milligram

   D liter

9. What is the ***best*** use for an MSDS sheet?

   A to determine how to correctly use a Bunsen burner

   B to determine the correct procedure for your lab experiment

   C to determine the best disposal method for a chemical

   D to determine the correct volume of chemical to use

10. Looking through a microscope, you made some observations about a sample. Which SI unit would be ***best*** to use to describe your observations?

    A km

    B m

    C L

    D mm

Chapter 1
Cellular Structure

This chapter covers:
1.1.1, 1.1.2, 1.1.3

THE CELL

The **cell** is the structural and functional unit of organisms. It is the basis for all life on Earth. Cells come in a variety of shapes and sizes. Some cells, like blood cells, are microscopic and can only be seen with a microscope. Figure 1.1 shows a red blood cell. Other cells, like egg cells, are quite large and can be easily seen with the naked eye. Some cells use structures called organelles to survive. **Organelles** are small, specialized cellular subunits that carry out specific functions within the cell. Often organelles are separated from other cellular parts by a membrane.

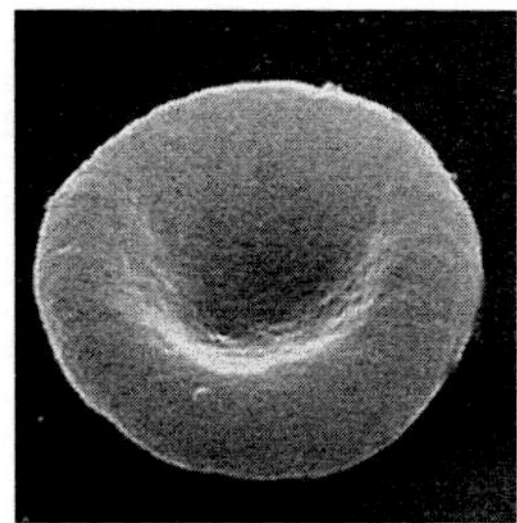

Figure 1.1 Blood Cell

Cellular shapes are as unique and varied as their functions. Think about your body — what do your skin cells do for you? Right! They surround and protect your body, and as such their shape is flat, durable and flexible. Nerve cells, on the other hand, transmit messages and are long, thin and very fragile. Ultimately, the function of the cell determines its size and shape. Figure 1.2 shows examples of skin cells, while Figure 1.3 shows a nerve cell. Note their different shapes.

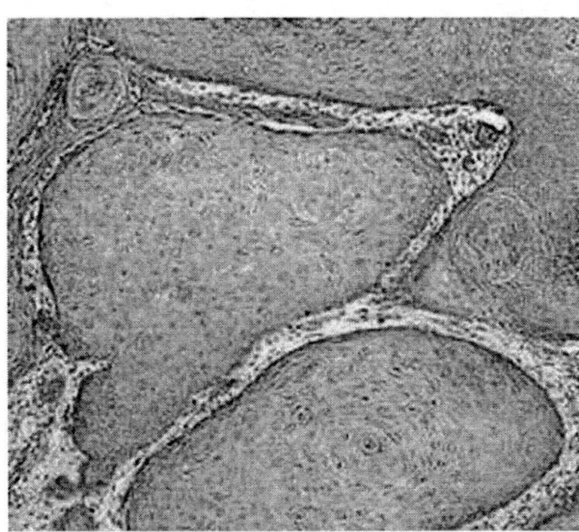

Figure 1.2 Skin Cell

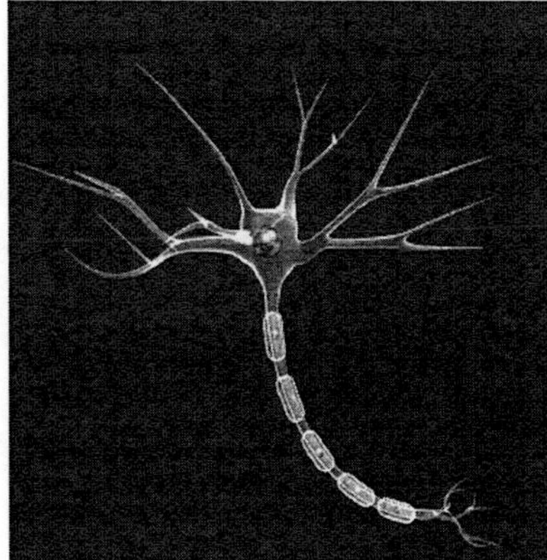

Figure 1.3 Nerve Cell

PROKARYOTIC VS. EUKARYOTIC CELLS

There are two basic types of cells: prokaryotic and eukaryotic. A **prokaryotic** cell does not have a true nucleus. Bacteria are prokaryotic. Although, the genetic material in a prokaryotic cell is usually contained in a central location; a membrane does not surround it. In prokaryotic cells, DNA is usually circular in structure and is called a plasmid. Furthermore, prokaryotic cells have no membrane-bound organelles. Generally speaking, prokaryotic cells are less complex than eukaryotic cells.

Prokaryote

Nucleoid

Capsule

Flagellum

Ribosomes

Cell membrane

Cell wall

Figure 1.4 Prokaryotic Cell

Bacterial cells are structurally different from eukaryotic cells. They often have specialized locations on or in their membrane where they carry out specific tasks or store fatty granules (food reserves). Lastly, the cell wall of most bacteria is structurally different from other cells. It is made from a special molecule called peptidoglycan. This molecule is a long chain made up of sugars and amino acids. It forms a net-like layer that surrounds the outer plasma membrane of the prokaryotic cell.

Eukaryote

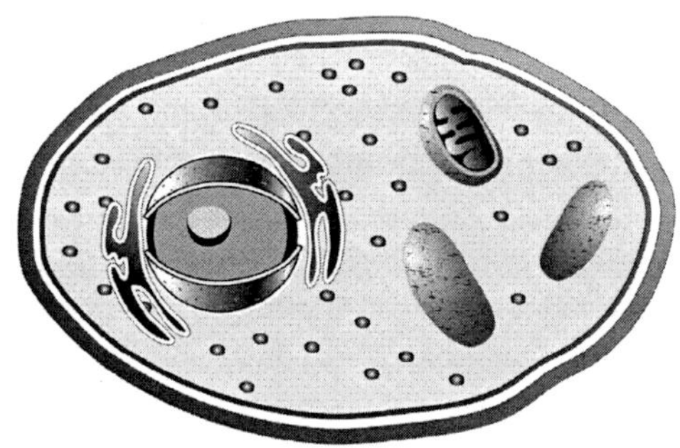

Figure 1.5 Eukaryotic Cell

A **eukaryotic** cell has a nucleus surrounded by a nuclear membrane. Genetic material in a eukaryotic cell is surrounded and protected by a nuclear membrane. In eukaryotic cells, the genetic material is spiral shaped and is called a **chromosome**. Eukaryotic cells also have several membrane-bound organelles. Eukaryotic cells tend to be larger than prokaryotic cells. Eukaryotic cells can also be quite complex. Some eukaryotic cells take on special functions and develop highly specialized structures. Plant and animal cells are both eukaryotic and, although similar in structure, contain unique cell parts. For instance, plant cells have a cell wall and chloroplasts, while animal cells have centrioles and some even have cilia and flagella. See Figures 1.5 and 1.6 for schematic drawings of eukaryotic cells, including plant and animal cells.

While they do share many differences, there are a few similarities between prokaryotic and eukaryotic cells. For example, both types of cells contain genetic material (DNA and RNA). Both types of cells also contain ribosomes. Lastly, prokaryotic and eukaryotic cells both represent living organisms.

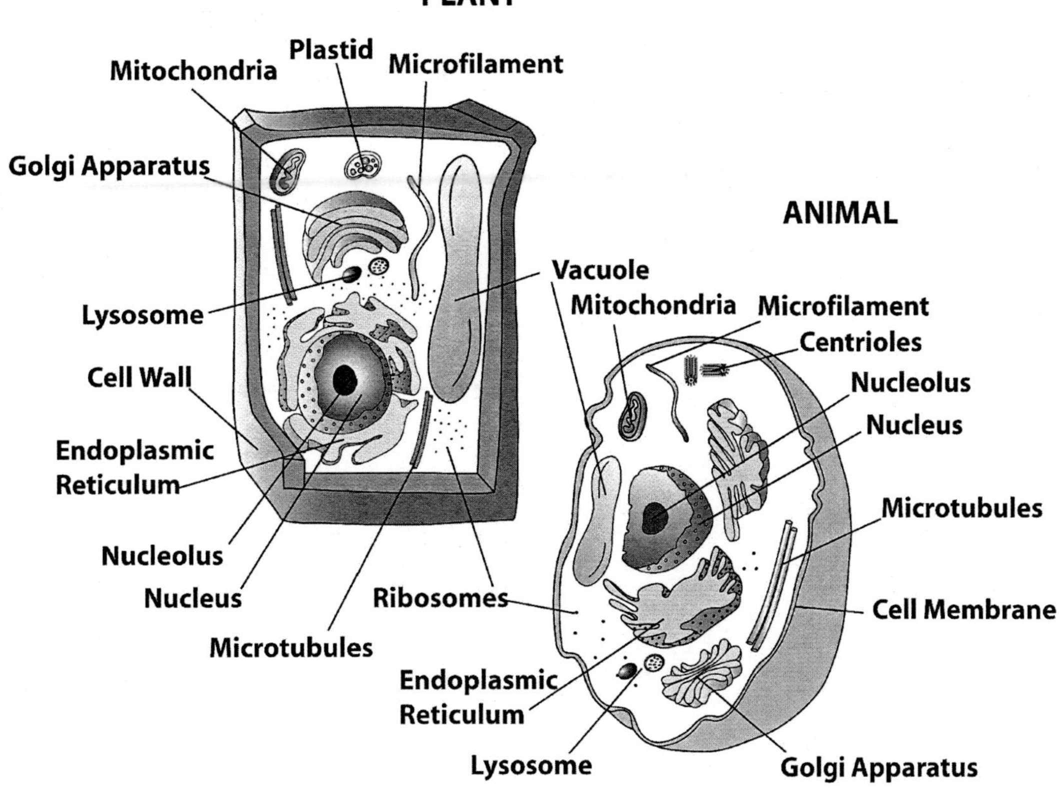

Figure 1.6 Parts of the Cell

Unicellular and Multicellular

Some cells can operate independently to survive. We call these types of cells free-living or **unicellular**. Algae is one example of a type of cell that can survive independently. Figure 1.7 shows a free-living organism called a volvox. Although the individual volvox cells can survive independently, they tend to congregate together and form a colony. The colony is much larger than the individual cells, increasing their chances for survival. Other cells must work together as part of a larger organism to survive. These types of cells are part of a **multicellular** organism. Humans are a multicellular organism.

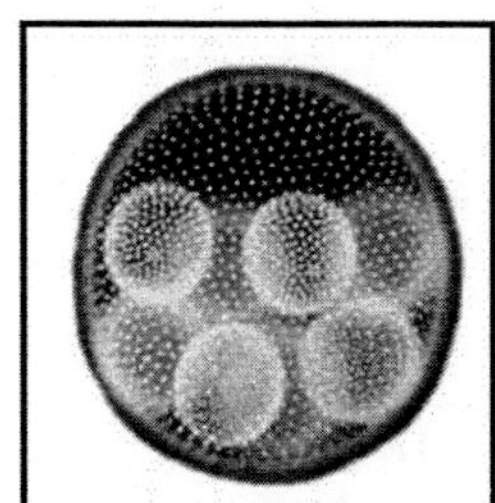

Figure 1.7 Volvox

Organelles

Organelles, or "little organs," are small, specialized cellular subunits that carry out specific functions within the cell. Organelles help to divide and organize the cell. They create specific locations where certain tasks can be accomplished. One example is waste removal; an organelle called a lysosome is responsible for removing waste from cells. Within the membrane of the lysosome, larger molecules are broken down or repackaged and removed from the cell. This compartmentalizing helps keep the cell efficient.

Cellular Structure

Among other things, organelles can help cells to move molecules, create and store energy, store information and perform a variety of other functions. Table 1.1 lists and describes some of the organelles and structures in cells.

Table 1.1 Parts of the Eukaryotic Cell

Name	Description
Cell Wall (plant cells, bacteria and fungi cells only)	rigid membrane around cell; provides shape and support
Plastids (plant cells only)	group of structures (chloroplasts, leukoplasts, chromoplasts) used in photosynthesis and product storage; have a double membrane and provide color and cellular energy
Vacuoles	spherical storage sacs for food and water
Cell Membrane	membrane surrounding the cell that allows some molecules to pass through
Golgi Apparatus	flattened membrane sac for synthesis, packaging and distribution
Mitochondria	rod-shaped double membranous structures where cellular respiration takes place
Microfilaments and Microtubules	fibers and tubes of protein that help move internal cell parts
Endoplasmic Reticulum (ER)	folded membranes having areas with and without ribosomes used for transport of RNA and proteins
Nucleolus	dense body in the nucleus; site of ribosome production
Nucleus	control center of the cell; location of hereditary information (chromosomes); surrounded by nuclear envelope
Nuclear Envelope	double membrane that surrounds the nucleus; fused at certain points to create nuclear pores; outer membrane is continuous with the ER.
Ribosomes	structures that manufacture proteins; found on endoplasmic reticulum and floating in the cytoplasm
Centrioles (animal cell only)	short tubes necessary for cell reproduction in some cells
Lysosomes	spherical sacs containing enzymes for digestive functions
Cilia (animal cell only)	short, hairlike extensions on the surface of some cells used for movement and food gathering
Flagella (animal cell only)	long, whiplike extension on the surface of some cells used for movement
Cytoplasm	jellylike substance in the cell that contains various organelles, inclusions and the cytosol
Cytosol	liquid like substance that makes up part of the cytoplasm. location of several metabolic processes, make up of water, proteins dissolved ions and other smaller molecules

Section Review 1: Cells

A. Define the following terms.

cell	plastids	nucleolus	tissue	vacuoles
organelles	Golgi apparatus	cilia	organ	cell membrane
cell theory	mitochondria	flagella	ribosomes	organ system
prokaryotic	microfilaments and microtubules	cytoplasm	centrioles	nucleus
eukaryotic	endoplasmic reticulum (ER)	unicellular	lysosome	
cell wall		multicellular		

B. Choose the best answer.

1 Which of the following is ***true*** about the mitochondrion of a cell?

A It has only one membrane.
C It is circular.
B It has no membrane.
D It is where cellular respiration occurs.

2 Which organism is ***most likely*** a unicellular organism?

A tree
C yeast
B human
D fish

3 What are structures that support and give shape to plant cells?

A microbodies
C nucleus
B Golgi apparatus
D cell walls

4 What word ***best*** describes the function of the organelle?

A incorporate
C unify
B join
D subdivide

5 Where is the hereditary information in a eukaryotic cell stored?

A cytoplasm
C centrioles
B nucleus
D lysosomes

C. Complete the following activity.

1 Compare and contrast prokaryotic and eukaryotic cells.

FERTILIZATION AND CELL DIFFERENTIATION

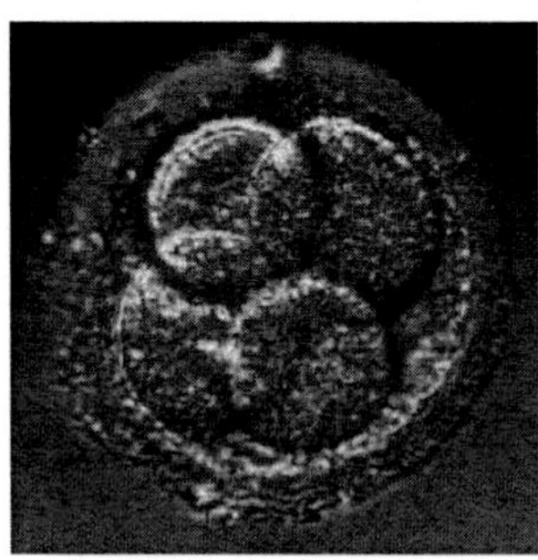

Figure 1.8 Zygote

During sexual reproduction, organisms produce sex cells called gametes. Females produce gametes called ova, and males produce gametes called spermatozoa. Gametes contain half the number of chromosomes needed to form a new organism. Gametes must come together in a process called fertilization. **Fertilization** occurs when the ova and spermatozoa fuse to form a new parent cell. When this happens, the new parent cell, called a **zygote**, is formed. The zygote has a complete set of chromosomes, half from the ova and half from the spermatozoa. It is a single cell capable of creating an entirely new organism. The newly formed zygote begins to grow. It first grows larger in size and eventually begins to divide to increase the total number of cells in the newly formed organism. When the zygote starts dividing, it is then called an **embryo**.

The group of cells produced in the very early stages of the embryo's growth are similar to the original zygote. They are called embryonic **stem cells**. Stem cells are a unique type of cell that can develop into any type of cell in the organism.

Eventually, when the embryo reaches 20 – 150 cells in size, this group begins to produce cells that are different from themselves. This process is called **cell differentiation**. This basically means that cells grow to look and function differently from one another. Directly following fertilization, cells in the developing zygote all look and function exactly the same. They divide by mitosis in the exact same way — two cells become four cells. Four cells become sixteen and so on. As development progresses, cells begin to look and function very differently. They take on different shapes and different functions. Eventually, embryonic cells become specialized and later begin to form complex tissues. As each cell differentiates, it uses RNA to produce proteins characteristic to its specific function. Within the zygote, these cells look very unique. They each have a distinct purpose. This is when cells begin to resemble their adult forms. Heart cells begin beating like a heart; eye cells begin to perceive light waves. Looking at a picture of a zygote with differentiated cells, you could easily tell the various cells apart, because they look very different. In Figure 1.9, you can clearly see the difference between differentiated and undifferentiated cells. In Stage 1 and Stage 2, all cells appear similar and have a uniform look; these are undifferentiated cells. In Stage 3 – 5, cells take on various shapes and functions; these cells are differentiated. Here, we can tell that the process of differentiation occurred between Stage 2 and Stage 3.

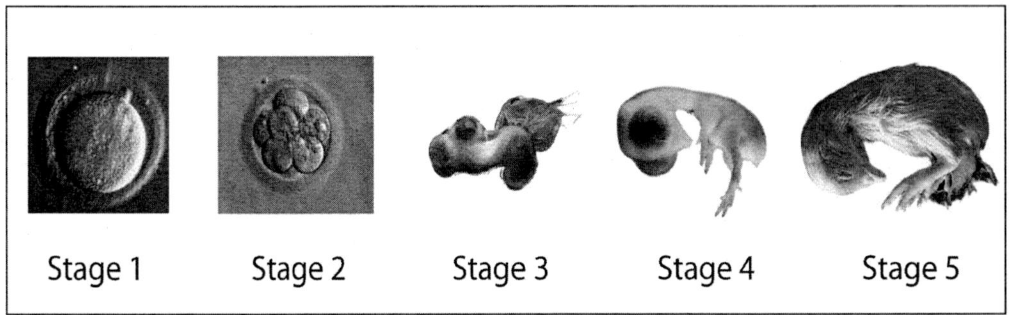

Figure 1.9 Cell Differentiation Stages of Zygote

Remember, stem cells have the capability to become any type of cell. This is possible because genes within the cell can be "turned on" or "turned off" at specific times. Every cell of the organism has the same genetic information that was present in the initial zygote. Thus, cell differentiation occurs by the selective activation or inactivation of only some of these genes. For example, some cells could become liver cells, while other cells become skin cells. Both of these cell types contain genes for every other cell type within the organism. But some of their genes were turned on or off at different times during development; therefore, the cells look and act completely different.

The mechanisms that control cell differentiation are complex. Often, environmental conditions can play a vital role in cell differentiation. Cells receive chemical signals from the environment and other cells that start the differentiation process. This is called **cell signaling**. While it is complicated, here is a basic synopsis of one type of signaling process. Cells in contact during embryonic growth produce varying amounts of a protein. The hyperproducing cells cause other cells nearby to reduce expression of certain membrane receptors, thus changing their shape and function. Because there was a large amount of the signaling protein in the environment, some cells produced fewer membrane receptors, beginning the differentiation process. This is another way organisms regulate gene expression.

If the environment surrounding these embryonic cells were not at the right temperature or did not have the correct pH, the signaling protein might break apart. Also, if the cells were not able to extract the correct nutrients from their environment, they might make a protein that is misshapen and unrecognizable to other cells.

Although most prominent during the embryonic stages of development, cell differentiation occurs throughout your lifetime. One example is when the body repairs itself after injury. New blood, bone or skin cells are grown from surviving lines of stem cells found within the body. It is important to remember that adult stem cells operate in the same way as embryonic stem cells. Under the right conditions, both embryonic and adult stem cells can develop into any cell type.

Section Review 2: Cell Differentiation

A. Define the following terms.

fertilization	cell signaling	stem cells	cell differentiation
zygote	embryo	asexual reproduction	stem cells

B. Choose the best answer.

1 In fertilization, what do gametes fuse to form?

A embryo

B somatic cell

C zygote

D reproductive cell

2 What are stem cells?

A cells that can produce any type of offspring cell

B cells that contain stem structures used in reproduction

C haploid cells that can produce any type of offspring cell

D cells which are found only in plants

Examples of Specialized Cells

As we have already learned, cells can differentiate and become physically different from one another. You might be asking, "Why do cells want to be different?" The answer is simple. They differentiate due to their specialized functions. As we mentioned at the beginning of the chapter, different cells perform different jobs for the organism. By dividing up the work to be done, cells increase their efficiency, making survival of the organism more likely. Both plants and animals have specialized cells. We will look at examples of specialized cells in plants and animals.

Specialized Cells in Plants

A plant is made up of three basic structural parts: roots, stems and leaves, all of which contain specialized cells. Plant cells grow from a region of the plant called the meristem or apical meristem. In the **meristem**, cells rapidly divide by mitosis to form new cells. Meristematic regions can be located at the tip of the root, shoot or along the stem of the plant. As the plant cells divide, they move through several zones until they reach the zone of differentiation, where they become mature. There are three basic types of cells found inside vascular plants: epidermal cells, ground cells and vascular cells.

Roots

Roots anchor the plant and help absorb water and minerals and store food. Roots grow from an apical meristem located near the tip of the root. At the very tip of the root is an umbrellalike covering called the **root cap**. The interior of the root cap is made up of living cells that divide rapidly. They develop thick cell walls and are quickly pushed outward toward the soil. Nearing the surface of the soil, they die and are stripped off the surface of the root cap. The sloughing off of these cells helps to penetrate the soil and protect the plant's other cells within the interior of the root. The cells of the root cap are epidermal cells.

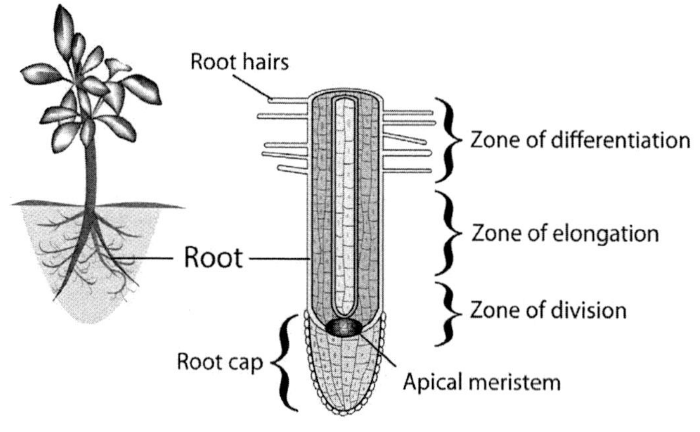

Figure 1.10 Plant Root Zone

The rest of the root's length is covered with a layer of thick-walled, tightly packed epidermal cells. In the zone of differentiation, these cells can develop small projections called **root hairs**. Root hairs increase the absorbent surface of the root.

The interior of the root is made up of a large area of parenchyma cells. **Parenchyma cells** are large similar cells that can be used to store nutrients. They remain alive in stasis and can function to repair the plant after injury. You can think of them as a kind of plant stem cell. The interior of the root is made up of another type of cell called **sclerenchyma cells**. These cells stack end-to-end inside the plant, eventually dying and forming a tubelike structure. The plant uses this tube to transport water from the roots to the leaves. Sclerenchyma cells are also sometimes called xylem cells. **Xylem** cells are vascular cells that transport water within the plant.

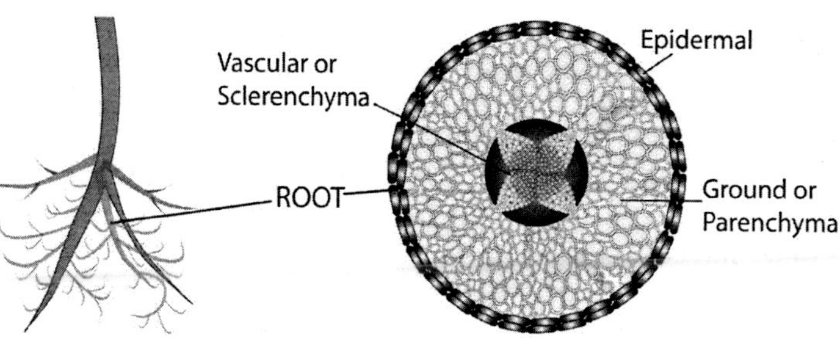

Figure 1.11 Cross Section of a Plant Root

STEMS

The stem provides support for the plant elevating its leaves, flowers and fruit. The stem transports nutrients from the roots to the leaves and vice versa. Sometimes in green fleshy plants, the stem performs photosynthesis, providing food for the plant. Stems can also store nutrients and grow new tissues. In land-dwelling plants, the stem has a waterproof covering made up of a layer of epidermal cells and a type of wax. The wax is called the cuticle. Some woody stems have cells called **cork** or **phellem**. Cork cells provide protection and prevent water loss. All stems contain vascular cells that transport water and nutrients throughout the plant. There are two main arrangements of cells inside a stem. Some stems have vascular cells located throughout their parenchyma region. Other stems have vascular cells located in a ring outside an innermost pith region.

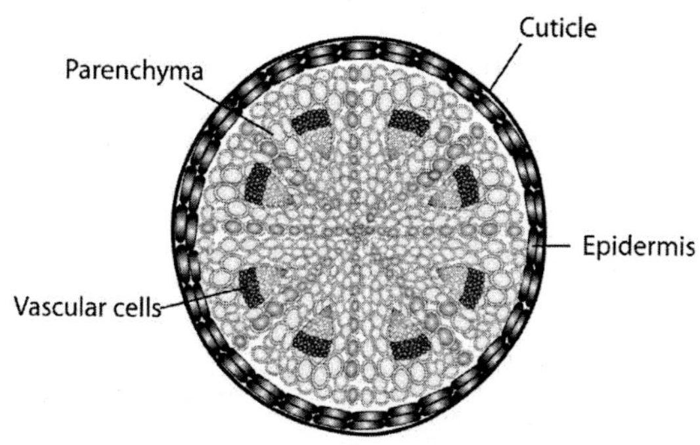

Figure 1.12 Cross Section of a Monocot Stem

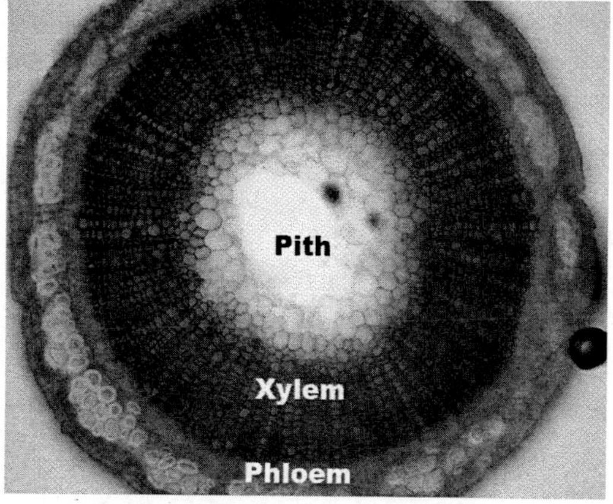

Figure 1.13 Cross Section of a Dicot Stem

LEAVES

All leaves must make food for the plant. There are many different types of leaves. On their exterior, plants have a waxy substance called a **cuticle.** Similar to the stem, the cuticle is secreted by the epidermal cells of the leaf. In dry or cold climates, plants have a thick cuticle to prevent water loss. Some plants have specialized hairs or **trichomes** located outside the cuticle.

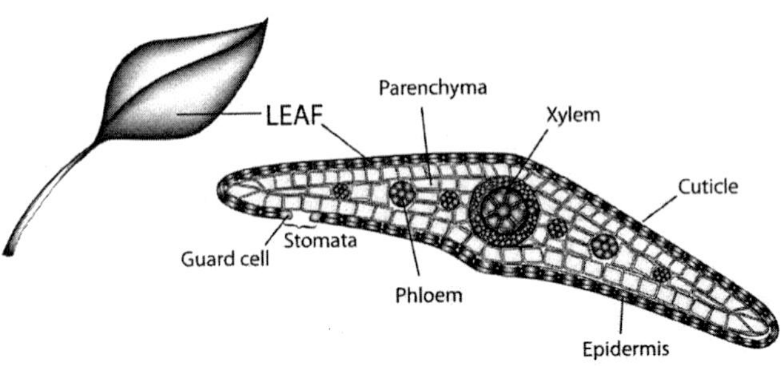

Figure 1.14 Leaf Plant Cross Section

Just inside the cuticle are the epidermal cells. Specialized cells called **guard cells** form openings called stoma. Stoma regulate gas exchange with the environment. Guard cells close during dry periods and reopen during wet periods. Leaves also contain parenchyma cells and vascular cells. Photosynthesis takes place in the parenchyma cells. Vascular cells called phloem cells transport food (glucose) from the leaf to other parts of the cell.

SPECIALIZED CELLS IN ANIMALS

Multicellular animals have a wide variety of cell types. Each type of cell functions to keep the organism living. We will look at three types of human cells to see specialization in animal cells.

BLOOD

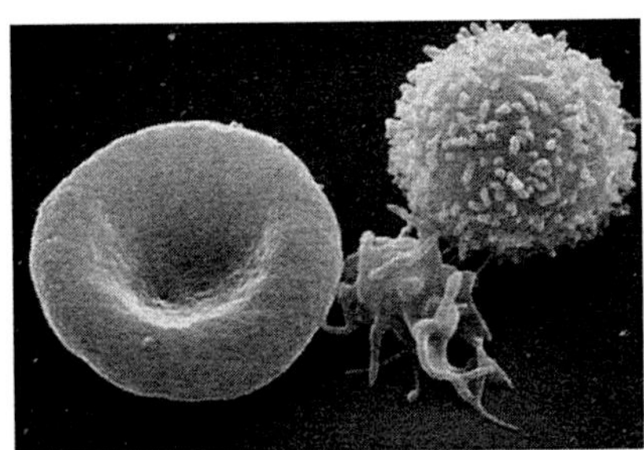

Figure 1.15 Three Types of Blood Cells

Blood is a fluid that transports substances throughout the body. It is made up of two main substances: plasma and blood cells. Plasma is over 90% water, but it is also made up of dissolved proteins, glucose, mineral ions, hormones and carbon dioxide. There are three main types of blood cells: red blood cells, white blood cells and platelets. Each cell type has a unique function that leads to distinctive forms.

Human **red blood cells** are disc-shaped cells that lack a nucleus and other organelles. Their primary function is to transport oxygen throughout the body. The disc shape increases their surface area, allowing each cell to transport lots of oxygen. They are primarily made up of hemoglobin — an iron-rich protein molecule that bonds well with oxygen. Red blood cells typically live about 120 days (4 months). Because they lack a nucleus, they cannot multiply by mitosis from other preexisting blood cells; instead they are produced from adult stem cells in the bone marrow.

Human **white blood cells** function to defend the body against disease. They can be found in the blood or lymph fluids. White blood cells come in a variety of shapes depending on their function. Some white blood cells have large granules that help them to consume and digest invading microbes. Others, like lymphocytes, have many proteins attached to their outer surfaces. These proteins or antibodies serve as a "memory" system enabling the organism to fight off future infections from that particular pathogen. They also give cells a "fuzzy ball" appearance.

Human **platelets** are also called thrombocytes. They form blood clots, thus slowing and stopping bleeding after injury. Platelets also lack a nucleus but contain powerful growth factors. Platelets stimulate cell growth surrounding the site of injury.

MUSCLE

Muscle cells shorten to cause movement of the body. In humans, muscle cells move the skeleton, heart and internal organs. The three main types of muscles are: skeletal, cardiac and smooth. Each type of muscle cell looks and functions differently.

Skeletal muscle cells are made up of bundles of long protein molecules called **myofilaments**. Many myofilaments bind together to form a fibril or **myofibril**. Many myofibrils are wrapped together by a cell membrane to form a long single muscle cell. One muscle cell can be more than one inch long. In between the myofibrils of the muscle cell are many, many, many mitochondria. They exist to provide the energy needed for muscle contraction. Within the cell membrane, on the outside of the cell, is the nucleus of muscle cells. A single muscle cell can have many nuclei — some have more than 500! Winding through the myofibrils is a system of specialized tubes that conveniently stores nutrients and conducts impulses between different muscle cells.

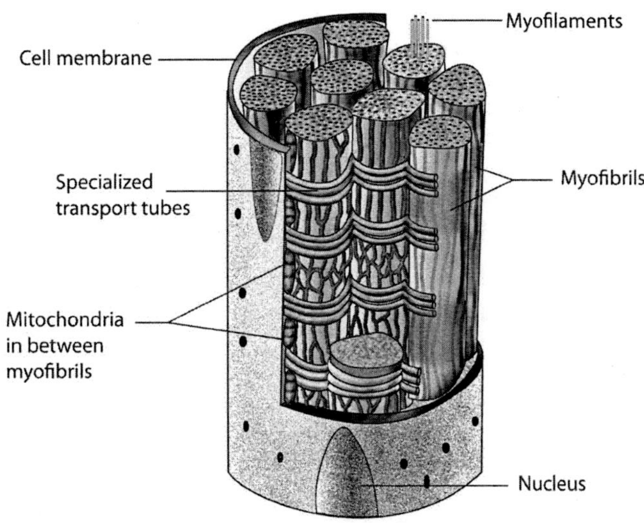

Figure 1.16 Structure of a Muscle Cell

To see some muscles in action, visit these web sites:

- http://multimedia.mcb.harvard.edu/anim_myosin.html (click on Myosin II)
- http://lessons.harveyproject.org/development/muscle/grsphysw.html

EPITHELIUM

Epithelial cells are cells that form the lining of structures. In humans, epithelial cells cover the outside of the body, creating a structure commonly called skin. These cells also cover the inside of body cavities, forming boundaries between different organs or organ systems. There are several different types of epithelial cells classified by their shape and arrangement. All epithelial cells are packed tightly together. The physical arrangement of epithelial cells provides a barrier, preventing many substances from passing through. Sometimes epithelial cells are layered together, providing a more imposing boundary.

Squamous cells are one type of epithelial cells — they are flat, thin and usually soft. They line the cheeks, blood vessels, lymph vessels, lungs and many other interior body cavities. **Cuboidal** cells, another type of epithelial cell, are cube shaped with their nucleus in the center of the cell. Lastly, **columnar** cells form a column shape, increasing the surface area of the space they cover. The small intestine is lined with columnar epithelial cells.

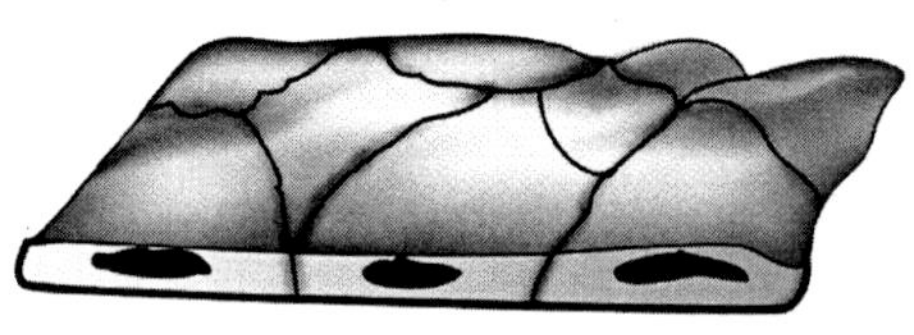

Simple Squamous Cell

Simple Cuboidal Cell

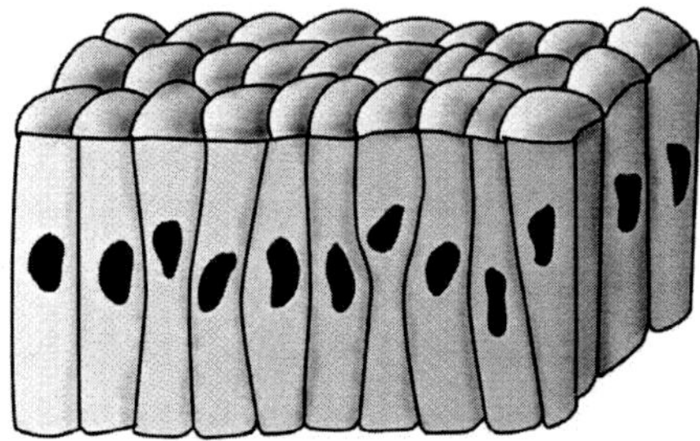

Simple Columnar Cell

Figure 1.17 Layers of Epithelial Cells

Section Review 3: Specialized Cells in Plants and Animals

A. Define the following terms.

meristem	white blood cells	myofilaments	squamous
root hairs	red blood cells	myofibril	cuboidal
root cap	parenchyma cells	epithelial	columnar
xylem	sclerenchyma cells	blood	guard cells
cuticle	cork or phellem	trichomes	platelets

B. Choose the best answer.

1 How does the structure of red blood cells show cellular specialization?

A It gives the cells the ability to pack tightly together with neighboring cells.

B The many mitochondria inside the cytoplasm allow it to use energy quickly.

C It is made up of long, tight bundles of protein that shorten to cause movement.

D The absence of a nucleus and round disc shape increases surface area.

2 How does the structure of plant leaves prevent water loss?

A Stoma open and close in response to cues from the plant.

B The parenchyma cells inside leaves store enough water for the plant.

C The sclerenchyma cells are large cells capable of moving lots of water.

D The meristem continually draws water from the soil.

3 What type of specialized cell is found in the woody stem of some plant cells?

A root cap B root hairs C cork D guard cells

4 Which type of cell has a nucleus embedded in its membrane?

A red blood cell C epithelial cell

B muscle cell D guard cell

C. Complete the following exercises.

1 Compare and contrast different types of blood cells.

2 Describe cells found in the root of a plant. How does a root's structure correlate with the harsh soil environment? How does the root's structure correlate with its function of water absorption?

CHAPTER 1 REVIEW

1 A ___________________ is a type of cell that has a ***true*** nucleus.

A prokaryote B eukaryote C bacterium D virus

2 What type of cell has a flagellum on its surface?

A an animal cell C a viral cell

B a plant cell D a diseased cell

3 What are two structures that are found in plant cells that are ***not*** found in animal cells?

A mitochondria and ribosomes

B cell wall and plastids

C cell membrane and centrioles

D nucleolus and endoplasmic reticulum

4 Which organelle is the site of protein synthesis?

A plastid C nucleolus

B ribosome D mitochondrion

5 Which cell part stores extra raw materials?

A vacuole C cell membrane

B nucleus D ribosome

6 Which cell part functions as storage for cellular information?

A vacuole B ribosome C nucleus D cell membrane

7 How does an endoplasmic reticulum function for cells?

A transports proteins C digests wastes

B stores water D makes DNA

8 Which cellular component is instrumental in maintaining homeostasis?

A ribosome C cell membrane

B cytoplasm D cilia

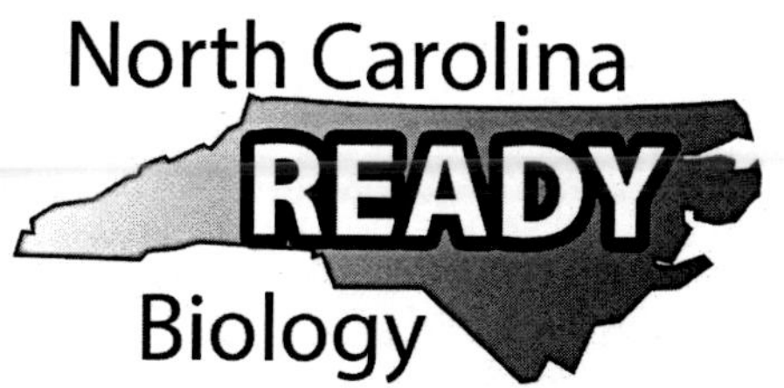

Chapter 2 Cellular Function

This chapter covers:
1.2.1, 1.2.2, 1.2.3

Homeostasis is Balance

Figure 2.1 Maintain Balance

If you recall, homeostasis is the maintenance of a stable internal environment. All living things must maintain a stable internal state to ensure survival. Toxins, microbes, availability of nutrients and temperature extremes can all threaten homeostasis. Luckily, organisms use many feedback mechanisms to maintain homeostasis. Internally, organisms use feedback systems to remove toxins, balance nutrients or stabilize temperatures. Externally, organisms use feedback mechanisms to respond to environmental changes. In this section, we will examine how organisms and cells respond to environmental changes and maintain balance in the system.

Internal Feedback of Organisms

Internal feedback systems are vital to maintaining life. Biochemical pathways constantly monitor and respond to changing internal conditions. These pathways are solely responsible for removing toxins, fighting foreign invaders, growth, digestion, reproduction or any other life-sustaining function.

In one example, increased exercise raises the level of carbon dioxide in the body. The body senses this increase and stimulates the respiratory system to increase the pace. The final result: heavy breathing. You have hundreds of internal feedback systems in your body. Each one is designed to keep the correct level of salts, fluids and nutrients available for your cells. Inside the human body, hormones are often used to maintain homeostasis. **Hormones** are chemical messengers inside your body that help transmit signals within the body. Hormones can help regulate internal body systems.

When you feel hungry or thirsty, it is your body's way of telling you to consume more food or drink. This is an example of an internal feedback mechanism. Can you think of yet another example of an internal feedback mechanism?

External Feedback of Organisms

Both plants and animals respond to their environments by using a feedback system. For example, when temperatures rise, most animals seek some type of shelter. They might move into the shade, or perhaps they dig a burrow. Animals respond to a variety of environmental stimuli. Humans jump at loud sounds, moths are attracted to light and bears move toward tasty-smelling food sources.

Plants also use feedback mechanisms. Changing light levels signal to plants the changing of seasons. When light levels decrease, many plants lose their leaves in preparation for winter. Organisms have developed many different kinds of adaptations to respond to changing external stimuli. **Adaptations** are special features that help organisms to survive. They can be physical responses or behavioral responses. There are thousands, if not millions, of different plant and animal adaptations on Earth. All adaptations increase an organism's chance for survival.

Homeostasis in Cells

Solutions

A **solution** is a liquid mixture of **solute** dissolved in **solvent**. Think of salt water, a solution in which salt (the solute) is dissolved in water (the solvent).

The interior of a cell is also a solution. The cytoplasm is a watery jelly like substance (the solvent) that contains a variety of substances, like salt and minerals (the solutes). Maintaining the concentration of solutes in the cytoplasm is critical to cell function — too much or too little of any component causes damage to the cell. Through a variety of mechanisms, the cell strives to maintain an ideal balance of solutes. The process is referred to as **homeostasis**. Often, cells must use energy to maintain homeostasis. Read on to discover how!

Cellular Homeostasis

To maintain homeostasis, multicellular organisms and unicellular organisms operate very differently. In a multicellular organism, special tissues or organs operate to maintain a balance. In humans, our kidneys work very hard to preserve the correct amount of salts in our bloodstream. In addition, blood-buffering systems act to uphold the correct pH and glucose levels. To sustain proper temperatures, sweat glands in our skin, fluids in our body and our muscles function to keep us at a relative 98.6 °F. When it gets hot, we sweat, utilizing our sweat glands and bodily fluids. When it gets cold, our muscles make us shiver to warm up. In colder climates, animals often have a higher level of fats in their tissues to insulate against the cold.

In unicellular organisms, special organelles often strive to maintain homeostasis. We will learn more about these types of specializations at the end of the chapter. To maintain the proper temperature, unicellular organisms can simply move to a better location.

The Cell Membrane and Cellular Transport

Cells strive to maintain the correct balance of nutrients, salts and biomolecules. They must also constantly remove waste products. Individual cells move fluids and nutrients in and out through the semipermeable cell membrane. They can move these materials by either passive or active transport mechanisms to maintain homeostasis. We will learn about active transport in Chapter 11. In the next section, we will discuss the structure of the cell membrane and passive transport mechanisms.

Cell Membrane

The main purpose of the cell membrane is to regulate the movement of materials into and out of the cell. The cell membrane is **semipermeable**, or selectively permeable, meaning that only certain substances can go through. Membranes can select molecules by size and charge. For example, large or charged molecules cannot pass. The cell membrane is an integral part of a cell's quest to maintain homeostasis. It is an important regulator that controls the flow of many solutes and solvents within cells.

The cell membrane is composed of a **phospholipid bilayer** as shown in Figure 2.2. Each phospholipid layer consists of **phosphate groups** (phosphorous bonded with oxygen) attached to two fatty acid (lipid) tails. The layers arrange themselves so that the phosphate heads are on the outer edges of the membrane, and the fatty acid tails compose the interior of the membrane. Globular proteins used for various functions, such as transporting substances through the membrane, are embedded in the cell membrane. The **phospholipids** are free to move around, allowing the membrane to stretch and change shape. Other molecules like cholesterol are important stabilizing molecules found in the phospholipid bilayer. They help keep the bilayer together.

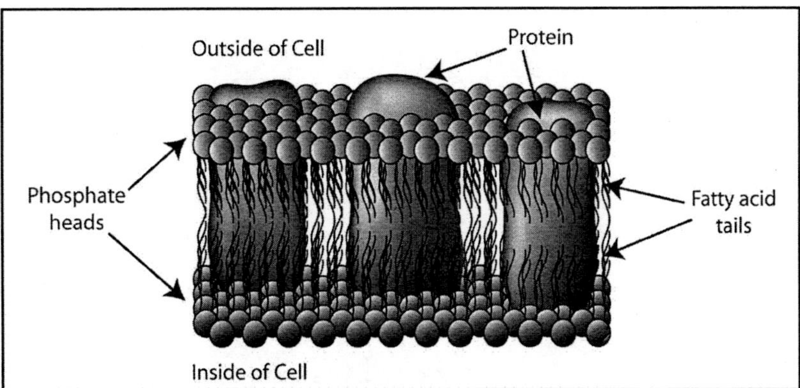

Figure 2.2 Phospholipid Bilayer

Passive Transport

Passive transport is spontaneous and does not require energy. In **passive transport**, molecules move spontaneously through the cell membrane from areas of higher concentration to areas of lower concentration; they are said to move "with the **concentration gradient**." The three types of passive transport are diffusion, facilitated diffusion and osmosis.

Diffusion is the process by which substances move directly through the cell membrane as shown in Figure 2.3. Osmosis is a type of simple diffusion. **Facilitated diffusion** uses a channel or carrier protein to move a substance from one side of the cell membrane to the other. Often large molecules or charged molecules use channel proteins.

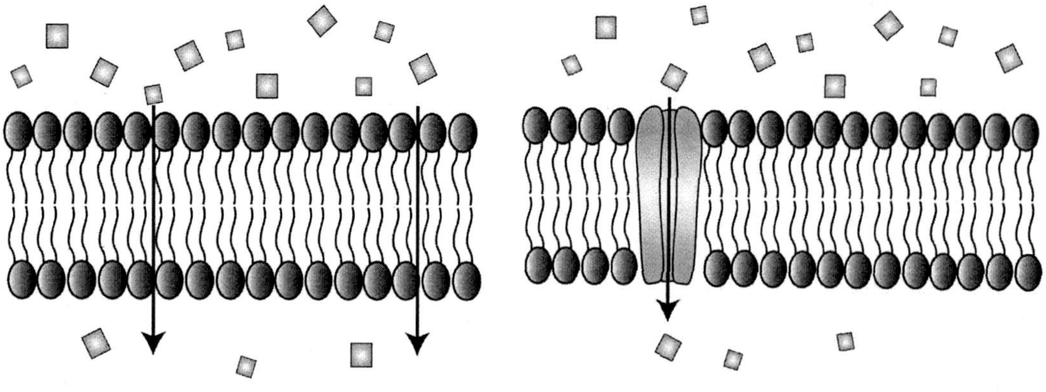

Simple

Facilitated

Figure 2.3 Diffusion

Osmosis is the movement of water from an area of high water concentration to an area of low water concentration through a semipermeable membrane. Figure 2.4 shows osmosis. Think of osmosis as the diffusion of water.

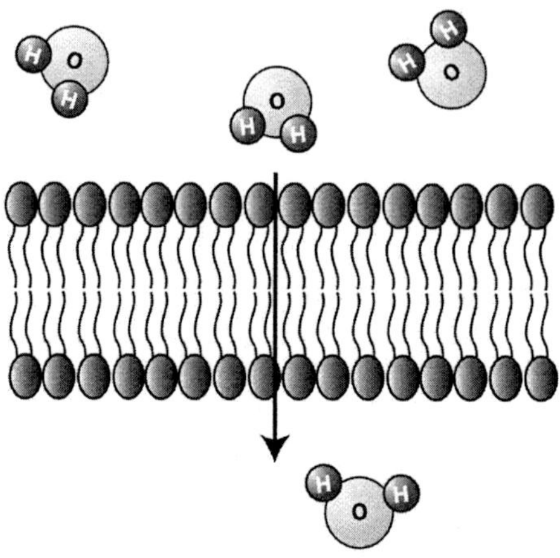

Figure 2.4 Osmosis

Osmosis can occur in either direction, depending on the concentration of dissolved material inside and outside the cell. Defining the solution concentrations *relative to one another* will predict the direction in which osmosis will occur. A **hypotonic** solution has the lower concentration of solute; this may be thought of as a higher concentration of water. A **hypertonic** solution has a higher concentration of dissolved solute, which may be thought of as a lower concentration of water. If the solute concentrates are the same inside and outside the cell membrane, the solutions are said to be **isotonic** to each other. Diffusion of water (osmosis) across a cell membrane always occurs from hypotonic to hypertonic. Three situations are possible and illustrated in Figure 2.5.

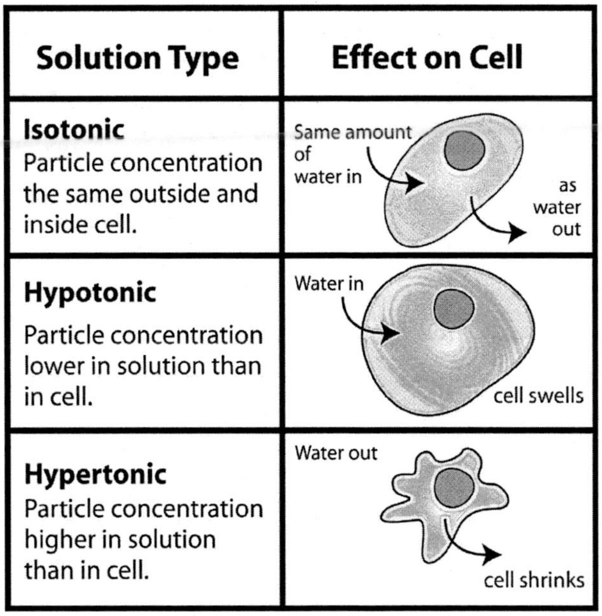

Solution Type	Effect on Cell
Isotonic Particle concentration the same outside and inside cell.	Same amount of water in as water out
Hypotonic Particle concentration lower in solution than in cell.	Water in cell swells
Hypertonic Particle concentration higher in solution than in cell.	Water out cell shrinks

Figure 2.5 Possible Results of Osmosis

Placing plant cells in a hypertonic solution causes the plant cell membranes to shrink away from the cell wall. Because there is more water inside the cell, water moves from the cell to the solution. This process is called **plasmolysis**. Plasmolysis can result in plant cell death due to water loss. A wilted plant is showing signs of plasmolysis. This is what can happen when plants are given too much fertilizer. Lots of fertilizer dissolved in solution means there is more water inside the plant cell than in the solution. This causes the plant to lose water and wilt. Placing a plant in a hypotonic solution has an opposite effect: The cell will swell until the cell wall allows no more expansion. The plant now becomes very stiff and turgid. Another term we need to be familiar with is osmotic pressure. **Osmotic pressure** is the pressure required to maintain equilibrium between two solutions separated by a semipermeable membrane. It is related to the difference in solute concentration between the two solutions. A large discrepancy in the solute concentration between the two solutions means a higher osmotic pressure. If two solutions have a similar solute concentration, they will experience a lower osmotic pressure. Put another way, a high concentration gradient means an increased osmotic pressure, while a low concentration gradient means decreased pressure. When cells are in a hypotonic environment, water moves into the cell via osmotic pressure thus causing the cell to expand. In plants, the cell wall prevents excessive swelling and creates turgor pressure. Turgor pressure allows plants to stand upright and regulate the movement of their stoma. In animals, too much osmotic pressure can cause the cell to swell until the membrane ruptures. This can kill the cell.

Kidney dialysis is an example of a medical procedure that involves diffusion. Another example is food preserved by salting, sugar curing or pickling. All of these examples are methods of drawing water out of the cells through osmosis.

Activity

Write a creative story or comic strip about a cell found in an isotonic, hypotonic or hypertonic solution. How did it survive? Earn extra points for creative new cellular adaptations.

Section Review 1: Homeostasis

A. Define the following terms.

diffusion	osmosis	homeostasis
facilitated diffusion	concentration gradient	hormones
osmotic pressure	passive transport	phospholipids
plasmolysis	solvent	semipermeable
hypotonic	solute	phospholipid bilayer
	hypertonic	phosphate groups

B. Choose the best answer.

1 Cells strive to maintain

A hormones.

B homeostasis.

C habitat.

D happiness.

2 Shipwreck survivors often attempt to drink sea water out of desperation. Predict what will ***most likely*** happen to survivors who drink sea water.

A They will become more hydrated as their cells swell.

B They will become more dehydrated as their cells shrink.

C They will become more dehydrated as their cells swell.

D They will become more dehydrated as their cells remain unchanged.

3 A cell has an internal salt concentration of 1.2 g/mL. If it is placed in a solution with a salt concentration of 0.12 g/mL, what will ***most likely*** happen to the cell?

A It will swell.

B It will shrink.

C It will remain unchanged.

D It will die.

4 Passive transport moves molecules

A with the concentration gradient.

B against the concentration gradient.

C toward the concentration gradient.

D away from the concentration gradient.

C. Complete the following exercises.

1 Draw a cell in a hypotonic solution.

2 Draw a cell in a hypertonic solution.

3 Draw a situation with high osmotic pressure and a situation with low osmotic pressure.

The Cell Cycle

The **cell cycle** is the sequence of stages through which a cell passes between one cell division and the next. The length of time it takes a cell to complete the cell cycle varies from one cell to another. Some cells complete the entire cycle in a few minutes, and other cells spend their entire life frozen in a particular phase.

Most of the cell cycle is spent in **interphase**, as shown in Figure 2.6. Interphase consists of three major parts: G_1, S and G_2. During the G_1 phase of interphase, the cell grows in size. In the S phase, replication of the DNA containing the genetic material occurs. In the G_2 phase, the cell prepares for mitosis by replicating organelles and increasing the amount of cytoplasm.

THE CELL CYCLE

INTERPHASE
G_1
Period of cell growth before the DNA is replicated (interphase begins in daughter cell)
S
Period when the DNA is replicated (that is, when chromosomes are replicated)
G_2
Period after DNA is replicated; cell prepares for division
MITOSIS
cytokinesis
telophase
anaphase
metaphase
prophase

Figure 2.6 The Cell Cycle

Mitosis

All of the cells in the body, with the exception of reproductive cells, are called **somatic cells**. Some examples are heart cells, liver cells and skin cells. Somatic cells undergo a process called mitosis. **Mitosis** is a type of cell division that generates two daughter cells. Both new cells are identical to the mother cell. This means mitosis is a form of asexual reproduction. During these phases, DNA is organized into thread-like material called chromosomes, then divided and separated.

The daughter cells that result from mitotic cell division are identical to each other as well as to the parent cell. The daughter cells have the same (diploid) number of chromosomes as the parent cell. Mitosis is the mechanism for **asexual reproduction**, which only requires one parent. Mitosis also allows multicellular organisms to grow and replace damaged or worn out cells. The stages of mitosis are: prophase, metaphase, anaphase and telophase. During **prophase**, DNA molecules condense into chromosomes. In **metaphase**, chromosomes line up in the middle of the cell. During **anaphase**, chromosomes separate and begin to move to opposite sides of the cell. Lastly, in **telophase**, the chromosomes arrive at opposite ends of the cell and the cell begins to divide.

Cell Division in Animal Cell

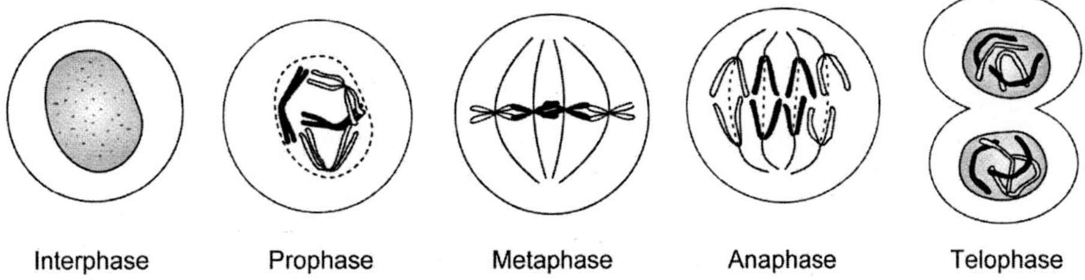

Figure 2.7 Mitosis in Animals

Cytokinesis

Cytokinesis, the division of the cell cytoplasm, usually follows mitosis. Cytokinesis generally begins during the telophase of mitosis. It finalizes the production of two new daughter cells, each with approximately half of the cytoplasm and organelles as well as one of the two nuclei formed during mitosis. The processes of mitosis and cytokinesis are together called **cell division**.

Section Review 2: Mitosis

A. Define the following terms.

mitosis	prophase	anaphase	cytokinesis
cell cycle	metaphase	telophase	interphase

B. Choose the best answer.

1 What process divides the cytoplasm of the cell?

A chromosomes
C mitosis
B cytokinesis
D chromatids

2. The length of time it takes for a cell to complete the cell cycle is

A around two hours.
C the same for each kind of cell.
B different for each cell.
D around two minutes.

3. When chromosomes line up in the middle of the cell, which phase is occurring?

A prophase
C anaphase
B metaphase
D telophase

4. Which situation described below would ***most likely*** require mitosis?

A a plant flowering to collect pollen and later create seeds
B male and female salmon spawning in a stream bed
C a person getting a cut on their finger
D a female producing unfertilized egg cells

C. Complete the following exercises.

1 Draw a cell that has two chromosomes going through the cell cycle.

2 Draw a cell that has eight chromosomes going through mitosis.

3 You can turn your drawings into a flip book to "see" mitosis in action.

UNICELLULAR ADAPTATIONS

You will remember from previous science courses that an adaptation is anything that helps an organism to survive. Adaptations can be physical structures or behaviors. In the next section, we will look at how various structures and behaviors help unicellular organisms survive. In particular, we will examine a unique kingdom of unicellular organisms called protists.

PLANT-LIKE PROTISTS

Plant-like protists are known as algae and may be unicellular or multicellular. Although algae come in different colors, all algae have chlorophyll-containing chloroplasts and can make their own food. We will look at one example of plant-like protist: *volvox*.

VOLVOX

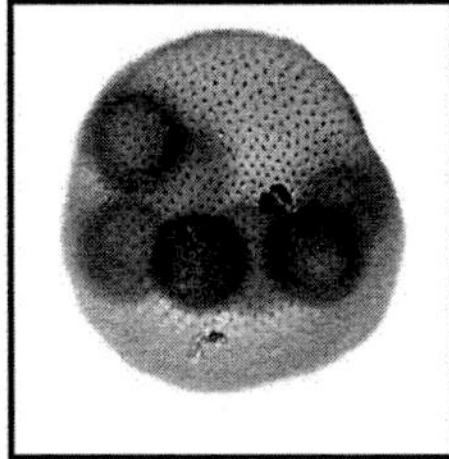

Figure 2.8 Volvox

Volvox is a type of green algae. Green algae store food in the form of glucose. They obtain their glucose by photosynthesis. Green algae use chlorophyll in photosynthesis to make glucose. The type of chlorophyll they use is very similar to all land plants.

Volvox is a colonial type of organism. It is made up of many individual flagellated cells arranged in a sphere. This means that if separated, each cell in a volvox can survive on its own. Volvox is important because it is generally considered to be the first evolutionary step toward a multicellular organism. Although each cell can survive on its own, it is more advantageous to remain within the colony. After all, a colony of 1,000 cells is much larger than a single cell. Actually, a colony of volvox (depending on the species) can be made up of anywhere from 500 – 50,000 cells.

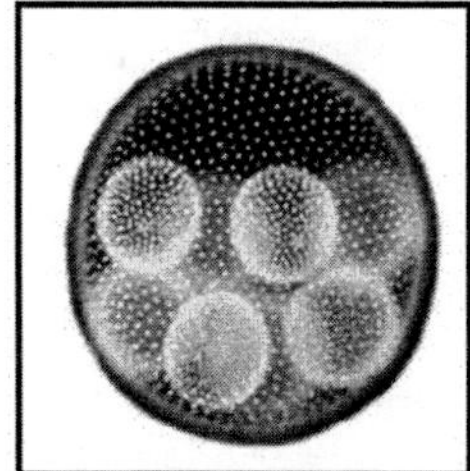

Figure 2.9 Volvox

You can find a colony of volvox living in freshwater puddles, lakes or ponds. They move by coordinating movement of their many flagella. Remember each cell in a volvox has flagella. They gently roll or tumble toward a light source. By coordinating the movement of the flagella, they can move the colony toward a light source and away from a harmful environment. Some of the cells in a volvox have an eyespot similar to a euglena. The cells that have an eyespot are usually located on one side of the colony. There is also new evidence to suggest that flagella help volvox gather more food. The movement of the flagella brings nutrient-rich water toward the colony and removes wastes. This allows more nutrients to diffuse into the interior of the colony. The cells in a volvox colony rely on osmosis, diffusion and exocytosis to remove waste products from their cells.

Cellular Function

Reproduction in a volvox colony is a little complicated. They can have both asexual and sexual reproduction. A specialized type of cell located in the volvox colony produces **daughter colonies**. This is a type of asexual reproduction. These cells use mitosis to repeatedly divide, making new colonies. The daughter colonies are often held inside the parent colony for a period of time. You can see the daughter colonies in Figure 2.10. They are the little balls inside the big one. A colony of volvox can also use sexual reproduction to create new colonies. Volvox can use special reproductive cells to make sperm or egg cells. This type of reproduction is often used when conditions become harmful, such as during a dry period. The specialization of reproductive cells seen in volvox is a common characteristic among multicellular organisms. This is yet another reason why volvox is considered an intermediate between unicellular and multicellular organisms. It is a common belief that green algae are the ancestors of all land plants.

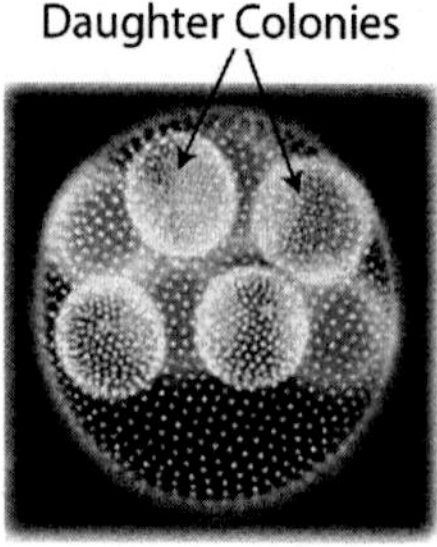

Figure 2.10 Volvox Daughter Colonies

Animal-like Protists

Animal-like protists are one-celled organisms known as protozoa. They must obtain their energy from other organisms and are called heterotrophs. Many protozoa are parasites living in water, on soil and on organisms. Many animal like protists respond to chemicals in the environment. the move toward food sources and away from waste products. This process is called **chemotaxis**.

We will look at three examples of animal-like protists: the euglena, amoeba and paramecium.

Euglena

Euglenas have characteristics of both plants and animals. They are both autotrophic and heterotrophic. This means they can survive solely by making food through photosynthesis, or they can hunt down and eat prey. Euglenas have chloroplasts that they use to capture light energy and make glucose. Or, when there is no light, they can survive by eating food. Food, like smaller algae or protists, is brought into the cell in a process called phagocytosis. **Phagocytosis** is a type of endocytosis where the cell membrane wraps around the food particle and forms a vacuole. You can think of it as a kind of "food vacuole." Smaller nutrient particles are brought into the cell using active transport or osmosis.

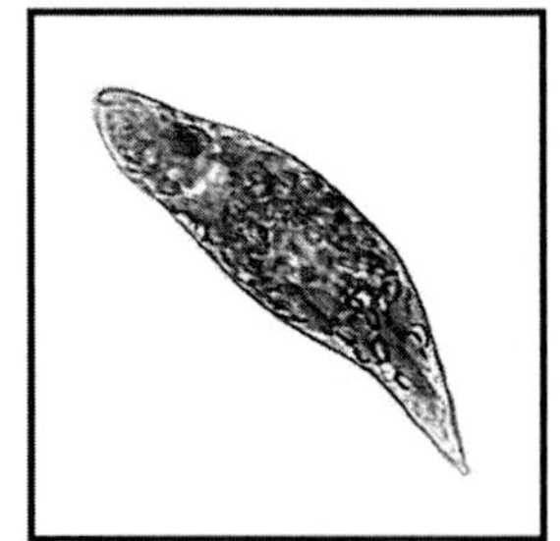

Figure 2.11 Euglena

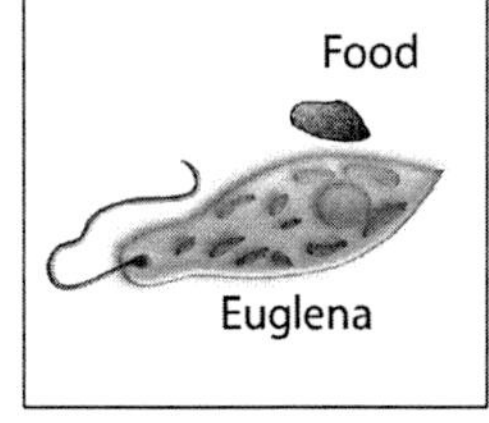

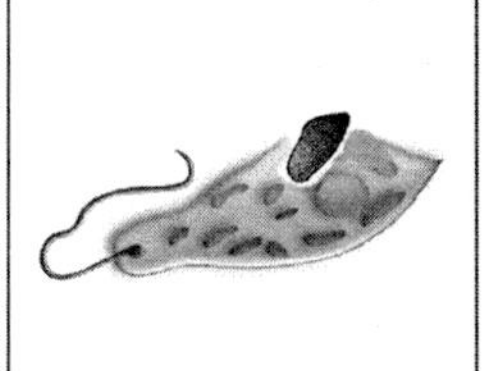

Figure 2.12 Phagocytosis

Euglenas live in fresh water, salt water or in the soil. When conditions are too tough for the euglena, it can form a protective covering, kind of like a spore. It then lays dormant and waits for things to improve. It commonly moves around in water with a long whip-like **flagellum**. A euglena can have one or more flagella.

The number of flagella is one identifying factor that can differentiate among various species. A euglena has no cell wall and is surrounded by a specialized type of cell membrane. It has an **eyespot** that responds to light. This eyespot is red in color, and it helps the euglena move toward light sources. Moving toward a light sources is called **phototaxis**. Euglenas reproduce using asexual reproduction, or mitosis. A single euglena can divide in half to become two new euglenas.

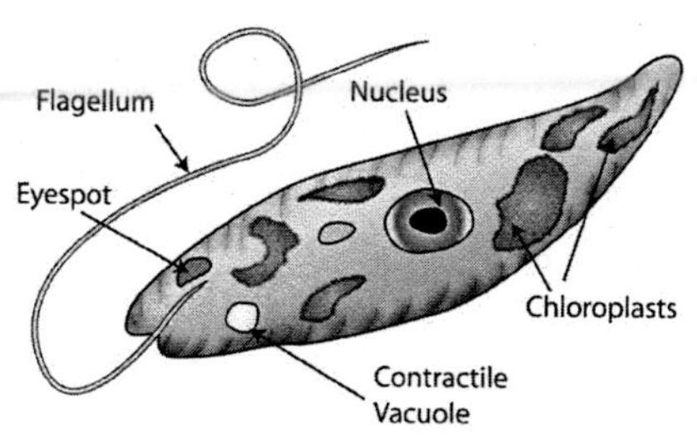

Figure 2.13 Euglena and Its Parts

A euglena uses a **contractile vacuole** to get rid of excess water. The vacuole is a storage sac that collects extra water from the cytoplasm of the euglena. Too much water in the cell causes the contractile vacuole to "squeeze" out the extra through an opening in the cell membrane. A euglena also uses lysosomes to dispose of other types of wastes. Exocytosis (the opposite of endocytosis) is used to expel wastes from the euglena.

PARAMECIUM

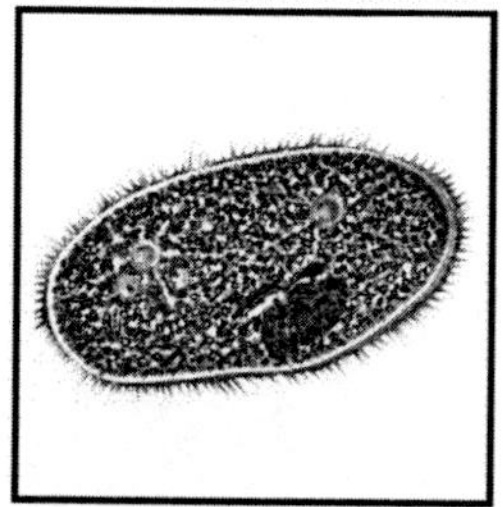

Figure 2.14 Paramecium

A paramecium is a typical ciliate. Ciliates have hairlike structures called **cilia**. Paramecia use cilia to move. They also use their cilia to help sweep food toward their "mouths." A paramecium eats bacteria, smaller protists and algae. It has a structure called an **oral groove**; this is like a mouth. Once food enters the oral groove, it is taken in by phagocytosis. This is similar to the way a euglena can ingest food. The food is then broken down by proteins that diffuse into the vacuole from the cytoplasm. The vacuole slowly shrinks in size. Once all the food has been digested, the vacuole moves toward the anal pore. Wastes are removed by exocytosis at the anal pore. A contractile vacuole helps to remove excess water inside a paramecium. Which other organism studied so far also has a contractile vacuole? Right, a euglena!

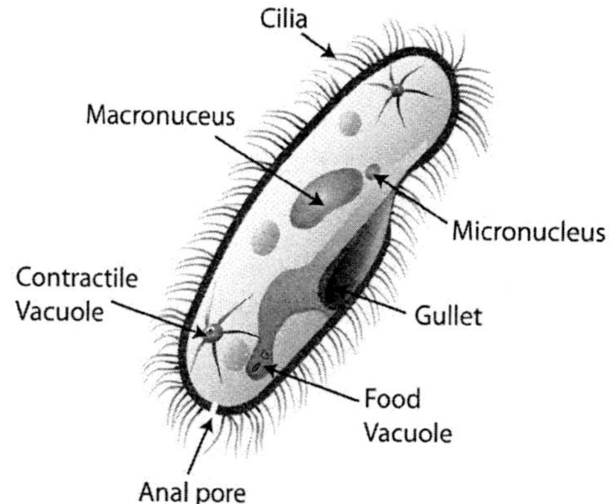

Figure 2.15 Paramecium Labeled

Paramecia reproduce asexually using mitosis. This is the most common form of reproduction. Ideally, paramecia can reproduce two or three times a day. Under stressful conditions, paramecia reproduce using sexual reproduction. They exchange genetic material in a specialized process called conjugation. Conjugation is their form of sexual reproduction.

Paramecia live in fresh and salt water.

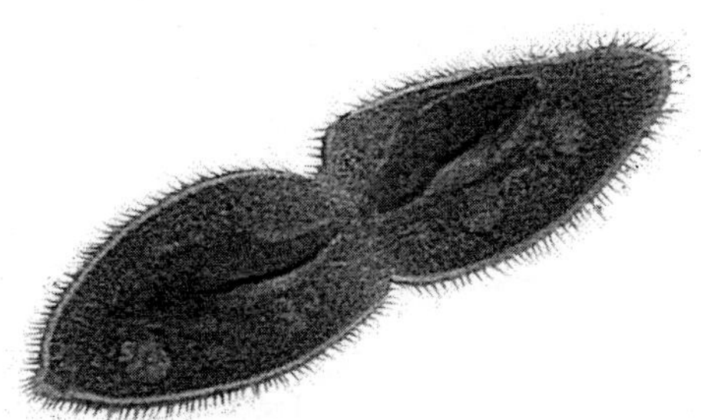

Figure 2.16 Paramecium Dividing

AMOEBA

Figure 2.17 Amoeba

An **amoeba** is a typical species of the Sarcodina phylum. Some species of amoeba (sometimes called sarcodines) can be quite large. They are often found in moist environments. Slow-moving freshwater ponds, lakes, streams and even marine environments can be good habitats for amoebas. Amoebas can also be found living on or inside animals (including humans)! Amoebas have no definite shape and are only surrounded by a cell membrane. They are often described as blobs. Amoebas move by **pseudopods**, which means false feet. This is also sometimes called amoeboid movement. They extend their blob-like feet to surround their food. Then, like euglenas and paramecia, they take in food particles using phagocytosis. Amoebas also use a contractile vacuole to get rid of excess water. Some amoebas have hard shells. When dead on the ocean floor, they form chalk and limestone. Amoebas reproduce asexually using mitosis.

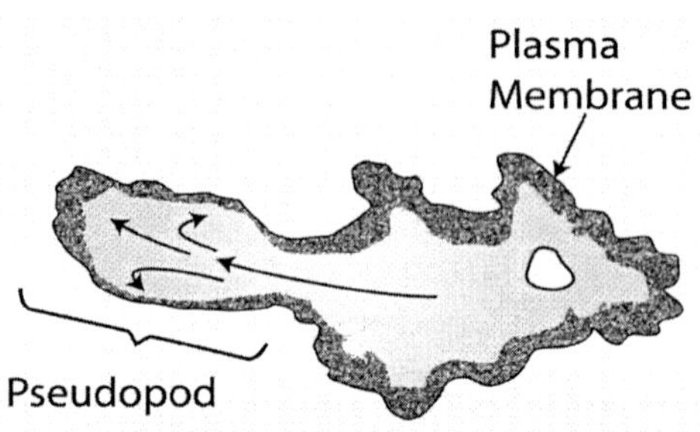

Figure 2.18 Amoeba Pseudopod

FUNGUS-LIKE PROTISTS

Fungus-like protists include several phyla that have features of both protists and fungi. They include slime molds, water molds and labyrinthulomycetes. They obtain energy from decomposing organic material. That's one reason they are considered a fungus. However, they are also able to move. That's what makes them a protist. Fungus-like protists move like an amoeba.

Fungus-like protists can be found in many environments including water, soils, forest floors, rotting wood or basements. Fungus-like protists often reproduce using spores, a structure often found in fungi.

Figure 2.19 Slime Mold

Section Review 3: Unicellular Adaptations

A. Define the following terms.

volvox	flagellum	chemotaxis
daughter colonies	eyespot	phototaxis
euglena	pseudopods	amoeba
phagocytosis	animal-like protists	fungus-like protists
plant-like protists		

B. Choose the best answer.

1 What is one way scientists differentiate among euglena species?

A number of daughter colonies
C number of flagella
B number of pseudopods
D number of cilia

2 Which two cellular structures help the euglena during phototaxis?

A membrane and contractile vacuole
B nucleus and ribosomes
C vacuole and cilia
D flagella and eyespot

3 Which cellular structure is vital to maintaining the correct balance of water and salts inside animal-like protists?

A nucleus
B eyespot
C contractile vacuole
D endoplasmic reticulum

C. Complete the following exercises.

1 Compare and contrast the reproductive strategies of the animal-like protists.

2 How are the animal-like protists different from plant-like protists?

Chapter 2 Review

1 More water goes in through a cell membrane than out of it. What type of solution is around the membrane?

A isotonic B hypotonic C hypertonic D permeable

2 Amoebas obtain food by wrapping the cell membrane around the food particle, creating a vesicle. The food is then brought into the cell. What is this process called?

A exocytosis
B endocytosis
C osmosis
D photosynthesis

3 A cell has an internal solute concentration of 11.4 g/mL. It is placed in a solution with a solute concentration of 12 g/mL. What will happen to the cell?

A It will swell.
B It will shrink.
C It will remain the same.
D It will become dehydrated.

4 A type of cell has several compartments surrounded by membranes. What type of cell must it be?

A prokaryotic
B semi selective
C eukaryotic
D bacterial

5 What type of molecule below ***most likely*** passes directly through the cell membrane?

A large glucose molecules
B charged potassium ions
C small water molecules
D large protein molecules

6 Organisms strive to maintain

A photosynthesis.
B homeostasis.
C osmosis.
D hormones.

7 When you perspire on a hot, humid day, drinking water will restore _______________ in your body.

A substances B oxygen C homeostasis D proteins

8 The cell membrane is

A solid. B semipermeable. C a solvent. D diffusion.

9 The water-saving thick skin of a cactus is a(n) ___________ characteristic.

A reproductive
B sensitive
C metabolic
D adaptive

10 A solution has a concentration of 0.63 g/mL. A plant cell has an internal concentration of 0.62 g/mL. What will ***most likely*** happen when the plant is placed in the solution?

A It will shrink.
B It will swell.
C It will remain unchanged.
D It will die.

11 A particular salt is in lower concentration inside the cell. This salt is required for cell metabolism. The cell must transport the salt from the exterior to the interior of the cell. What type of transport will the cell ***most likely*** use?

A passive
B active
C osmosis
D hypotonic

12 A cell has an internal concentration of 1.24 g/mL. Which solution below would create the highest osmotic pressure?

A Solution A at 1.25 g/mL
B Solution B at 1.00 g/mL
C Solution C at 2.0 g/mL
D Solution D at 5.00 g/mL

The image below shows the simplified process of osmosis.

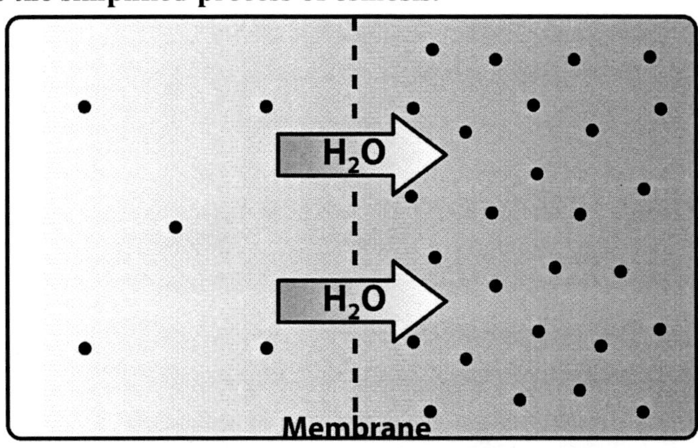

13 Which term ***best*** describes the solution on the left?

A hypotonic
B hypertonic
C isotonic
D cofactor

14 Which solution would create the lowest osmotic pressure for the cell below?

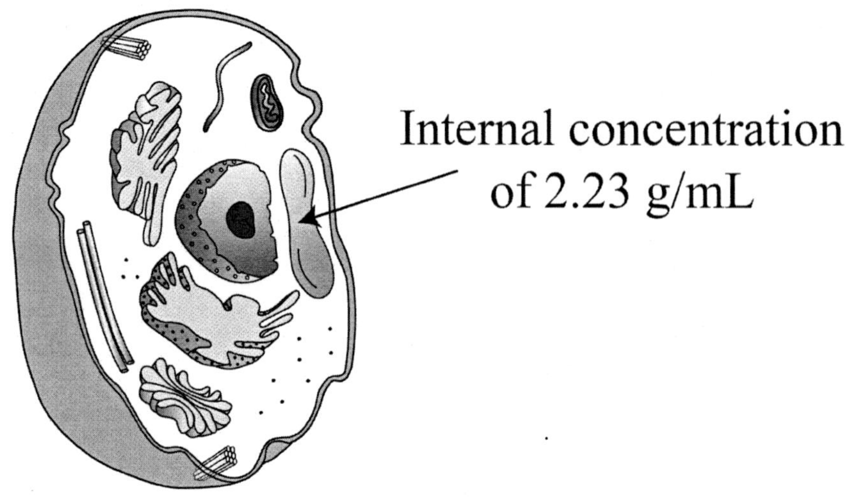

A Solution 1 at 2.00 g/mL

B Solution 2 at 1.25 g/mL

C Solution 3 at 5.00 g/mL

D Solution 4 at 2.25 g/mL

15 Which situation ***best*** represents homeostasis?

A condensation on a glass

B a monkey climbing a tree

C an alligator basking in sunlight

D a beaver building a dam

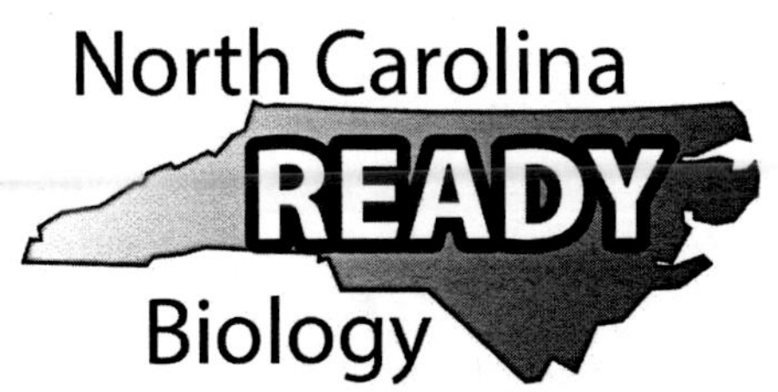

Chapter 3
Ecosystems

This chapter covers:
2.1.1, 2.1.2, 2.1.3, 2.1.4

The Nutrient Cycles

The process of recycling substances necessary for life is called a **nutrient cycle**. Nutrient cycles include the **carbon cycle**, the **nitrogen cycle** and the **water cycle**. When examining these cycles, it is important to remember that the elements are cycled in the real world. The nutrients shown in these images are not cycled as solo elements, but rather they often combine with other elements during their cycle. That is to say, when carbon and nitrogen cycle through the lithosphere, atmosphere or hydrosphere, they often carry other nutrients, like hydrogen or oxygen, along with them as ions that form salts. The images here show the cycles as stand-alone nutrient cycles for your learning benefit.

The important thing to remember is these nutrients are an integral part of the ecosystem. All parts including the plants, animals, rocks, water and air are connected. They help keep the ecosystem functioning. The chemical elements discussed in the chapter are used by plants and animals to form tissues, repair injury and grow. Constantly cycling these nutrients through the ecosystem provides a ready supply of nutrients to living things. Without these necessary nutrients, living things could not function.

The elements mentioned in this chapter move through the soil, sky and water systems in predictable ways. Natural processes on Earth help these atoms cycle through Earth's ecosystems. As a simple example, let's think about a molecule of oxygen. You probably know that trees give off oxygen — you might also know that animals breathe oxygen. So the really simple cycle might be that of plants releasing oxygen into the atmosphere. Animals take in oxygen from the atmosphere. Weathering of rocks can also release oxygen. Figure 3.1 shows this simplistic cycle. Plants and animals need oxygen for respiration, an energy-gathering technique.

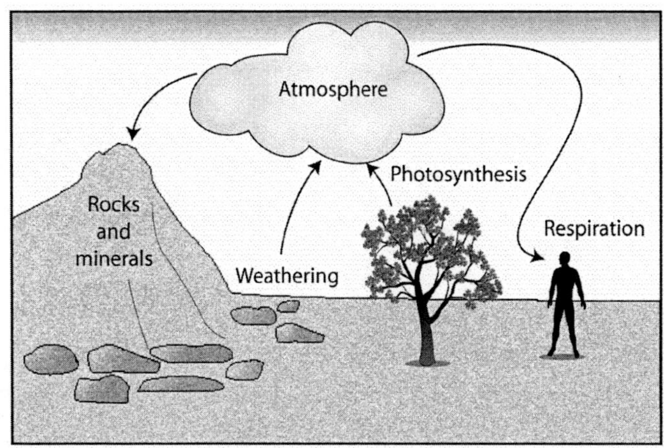

Figure 3.1 Oxygen Cycle

Carbon Cycle

The **carbon cycle** is the cycling of carbon between carbon dioxide (CO_2) and organic molecules. Organic molecules contain carbon – hydrogen bonds (C–H bonds). Inorganic carbon makes up 0.03% of the atmosphere as CO_2. Plants use CO_2 and energy from the Sun to perform photosynthesis. When animals eat plants, carbon passes into their tissues. Through food chains, carbon passes from one organism to another, as shown in Figure 3.2. Carbon returns to Earth through respiration, excretion or decomposition after death. Some animals do not decompose after death; instead, their bodies become buried and compressed underground. Over long periods of time, fossil fuels such as coal, oil and natural gas develop from decomposing organic matter. When fossil fuels burn, carbon dioxide returns to the atmosphere. This is one way humans contribute to climate change. Carbon is an important element for living things. Carbon's unique ability to bond to other elements makes it a primary component in most living things. The carbon cycle ensures a steady supply of elemental carbon for living organisms. Carbon is vital for photosynthesis and respiration.

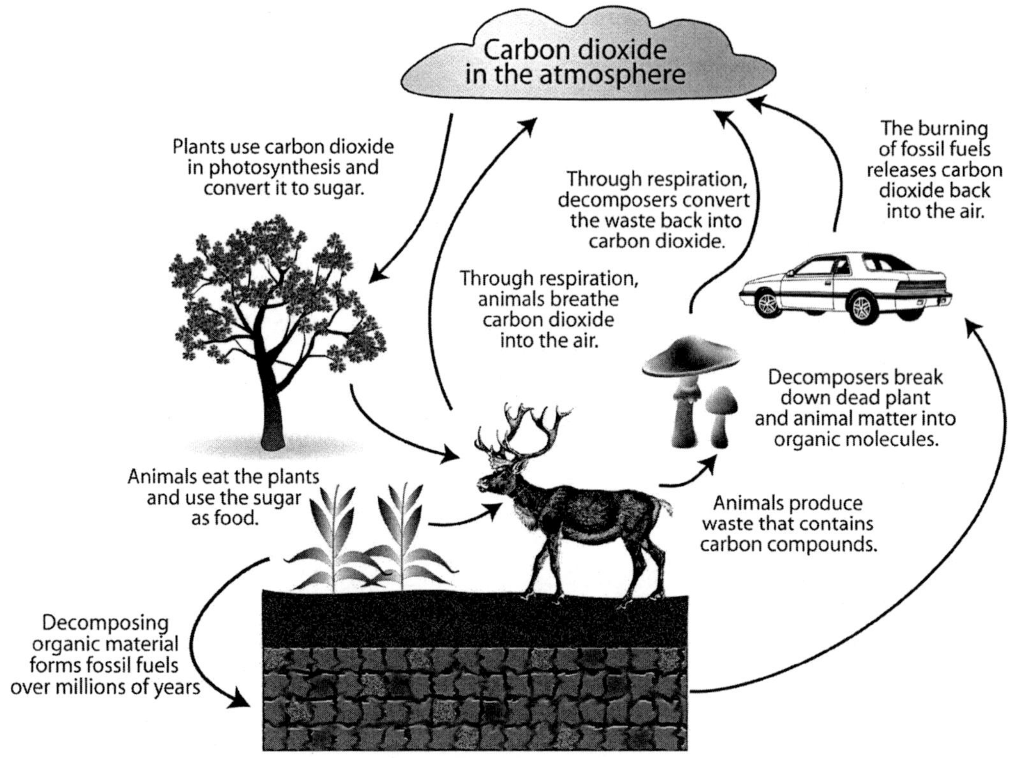

Figure 3.2 The Carbon Cycle

NITROGEN CYCLE

Nitrogen is the most abundant atmospheric gas, comprising 78% of the Earth's atmosphere. In fact, most of Earth's nitrogen is found in the atmosphere. However, nitrogen gas is not in a form that is usable by most organisms. The **nitrogen cycle** transforms nitrogen into ammonia, nitrite and finally nitrate, so that it is usable by plants and animals. Refer to Figure 3.3 to see the nitrogen cycle.

Nitrogen fixation is the conversion of nitrogen gas into nitrate by several types of bacteria. Nitrogen fixation occurs in three major steps. First, nitrogen is converted into ammonia (NH_3) by bacteria called **nitrogen fixers**. Some plants can use ammonia directly, but most require nitrate. **Nitrifying bacteria** convert ammonia into nitrite (NO_2^-) and finally into nitrate (NO_3^-). The bacteria live on the roots of legumes (pea and bean plants). This process increases the amount of usable nitrogen in the soil. The plants use the nitrogen, in the form of nitrate, to synthesize nucleic acids and proteins. The nitrogen then passes along through food chains. Decomposers release ammonia as they break down plant and animal remains, which may then undergo the conversion into nitrite and nitrate by nitrifying bacteria. Other types of bacteria convert nitrate and nitrite into nitrogen gas that then returns to the atmosphere. The nitrogen cycle keeps the level of usable nitrogen in the soil fairly constant. This makes nitrogen readily available to many plants. You might recall that DNA and RNA are made up of nitrogen. Thus, nitrogen is an important part of genetic structure for living things. Nitrogen is also a vital component of many amino acids, making it an important molecule during protein synthesis. A small amount of nitrate cycles through the atmosphere; this is created when lightning converts atmospheric nitrogen into nitrate.

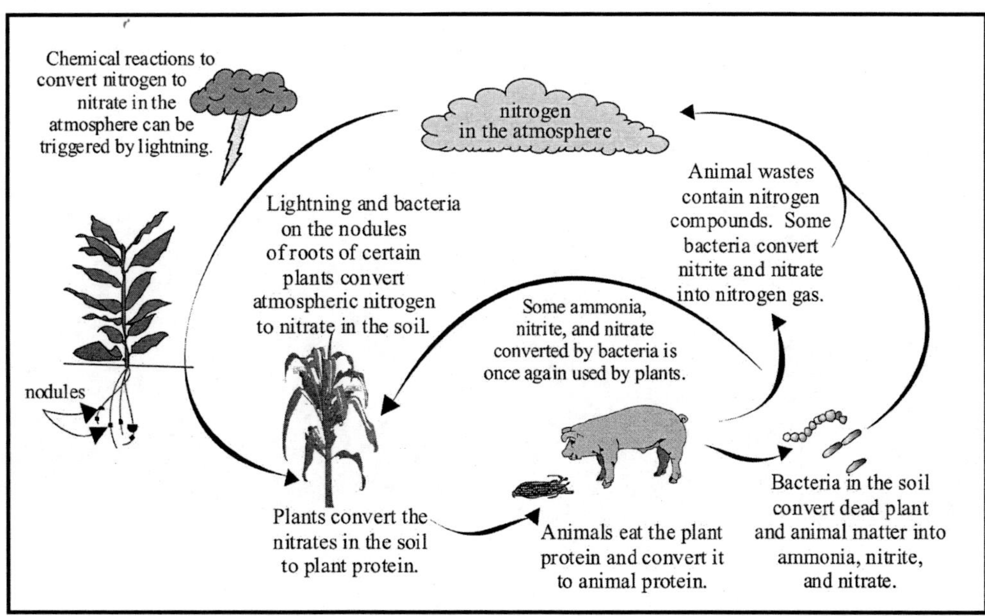

Figure 3.3 The Nitrogen Cycle

Challenge Question
Predict ways humans have impacted the nitrogen cycle.

WATER CYCLE

The **water cycle** circulates fresh water between the atmosphere and the Earth as seen in Figure 3.4. Even though water covers the majority of Earth, about 95% of it is salt water. Most of the fresh water is in the form of glaciers, leaving a very small amount of fresh water available for land organisms. Fresh water is vital for carrying out metabolic processes. The water cycle ensures that the supply of fresh water is available to plants and animals. Plants and animals require fresh water for respiration. Precipitation in the form of rain, ice, snow, hail or dew falls to the Earth and ends up in lakes, rivers and oceans through the precipitation itself or through runoff. The Sun provides energy in the form of heat, thus driving evaporation which sends water vapor into the atmosphere from bodies of water. Energy from the Sun also powers winds and ocean currents. Respiration from people and animals and transpiration from plants also sends water vapor to the atmosphere. The water vapor cools to form clouds. The clouds cool, become saturated and form precipitation. Precipitation can be in the form of rain, snow or hail. Fog or dew is another type of precipitation that returns gaseous water vapor to Earth. Without this cycle of precipitation, runoff and evaporation, a fresh water supply would not be available.

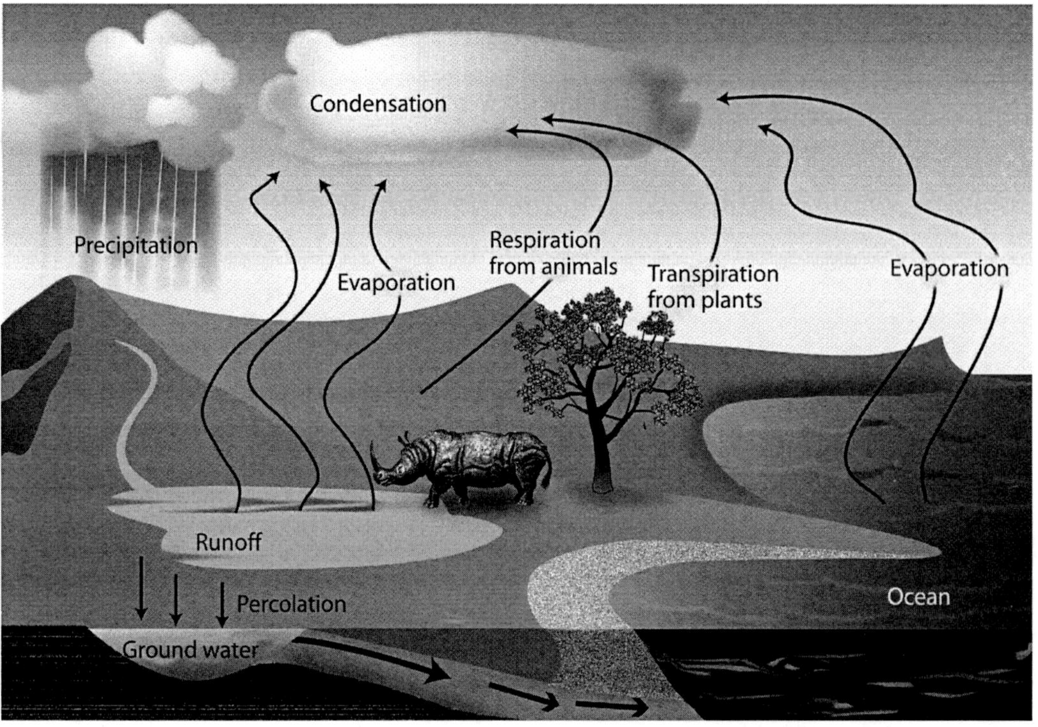

Figure 3.4 The Water Cycle

Section Review 1: The Nutrient Cycles

A. Define the following terms.

nutrient cycle	nitrogen cycle	nitrogen fixers	water cycle
carbon cycle	nitrogen fixation	nitrifying bacteria	

B. Choose the best answer.

1 Nitrogen-fixing bacteria perform which of the following tasks?

A They convert nitrogen to ammonia.

B They convert nitrogen to animal protein.

C They convert ammonia to plant protein.

D They convert nitrogen to plant protein.

2 Metabolic processes depend on which factor listed below?

A ice

B atmospheric nitrogen

C fresh water

D phosphorus

3 What is the main component of organic molecules?

A phosphorus

B carbon

C nitrogen

D carbon dioxide

4 How do plants use nitrogen?

A to make sugar

B to attract pollinators

C to make proteins and nucleic acids

D to transport water to their leaves

5 What is the name of the process by which water is transferred to the atmosphere by plants and trees?

A evaporation

B respiration

C condensation

D transpiration

6 Which of the following compounds is an organic compound?

A NH_3 B CH_4 C NO_3^- D H_2O

Energy Flow Through the Ecosystem

Matter within an ecosystem is recycled over and over again. Earth has the same amount of abiotic (not living) matter today as it did one hundred years ago. Elements, chemical compounds and other sources of matter pass from one state to another through the ecosystem. Recall from the last section the nutrient cycles, carbon cycles between living organisms and their environment. When matter is in living form, it is considered biomass. **Biomass** is a measurement of the amount of biological matter in an ecosystem.

As a deer eats grass, the nutrients contained in the grass are broken down into their chemical components and then rearranged to become living deer tissues. Waste products are produced in the deer's digestive system and pass from the deer's body back into the ecosystem. Organisms break down this waste into simpler chemical components. The grass growing close by is able to take up those components and rearrange them back into grass tissues. Then, the energy cycle begins again.

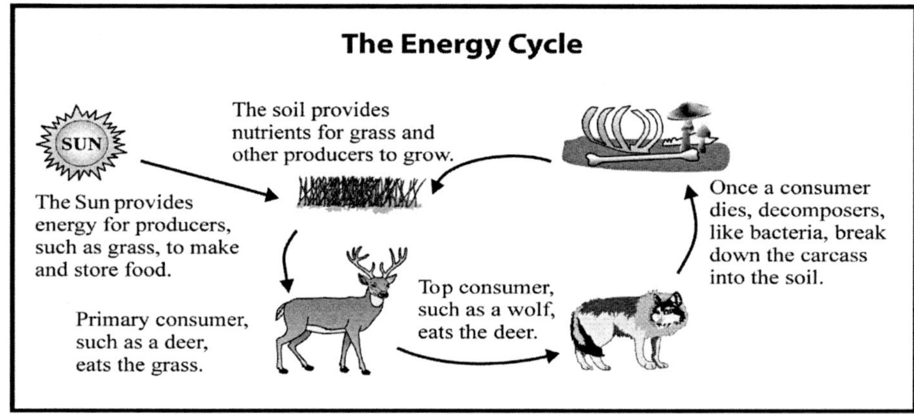

Figure 3.5 The Energy Cycle

Energy can be added, stored, transferred and lost throughout an ecosystem. **Energy flow** is the transfer of energy within an ecosystem. Inorganic nutrients are recycled through the ecosystem, but energy cannot be recycled. Ultimately, energy is lost as heat. Remember, however, that energy cannot be destroyed; although it may be lost from one system as heat, it is gained somewhere else. In this way, energy within any given system is conserved.

Food Chains and Food Webs

One way to graphically illustrate the transfer of energy within an ecosystem is a **food chain**. A food chain shows the connections between organisms, using arrows that point in the direction of biomass transfer. Another way to say this is that the arrows show who eats what (or whom)! Here we will look more closely at the interaction between producers, consumers and decomposers. From energy's point of view, we can say that energy generally transfers from producers to consumers to decomposers.

The **producers** of an ecosystem use abiotic factors to obtain and store energy for themselves or the consumers that eat them. In a forest ecosystem, the producers are trees, bushes, shrubs, small plants, grass and moss.

The **consumers** are members of the ecosystem that depend on other members for food. Each time a plant or animal consumes another organism, energy transfers to the consumer. Deer, foxes, rabbits, raccoons, owls, hawks, snakes, mice, spiders and insects are examples of consumers in a forest ecosystem. There are three types of consumers: **herbivores**, **carnivores** and **omnivores**. Table 3.1 lists characteristics of the three different types of consumers.

The **decomposers** are members of the ecosystem that live on dead or decaying organisms and reduce them to their simplest forms. They use the decomposition products as a source of energy. Decomposers include fungi and bacteria. They are also called **saprophytes**.

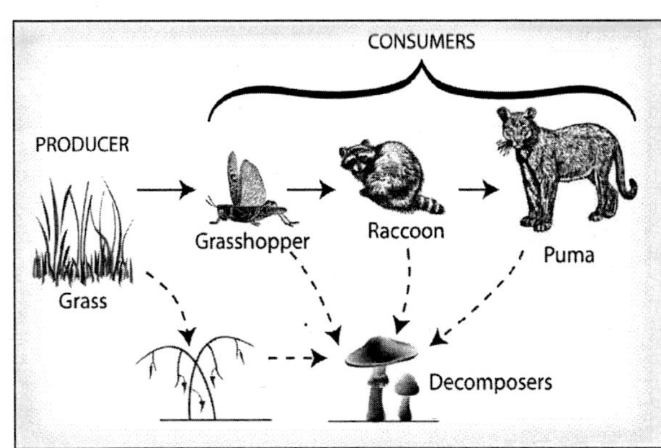

Figure 3.6 A Food Chain

Table 3.1 Types of Consumers

Consumer	Food Supply
Herbivore	animals that eat only plants
Omnivore	animals that eat both plants and other animals
Carnivore	animals that eat only other animals
Saprophytes	organisms that obtain food from dead organisms or from the waste products of living organisms

There can be many food chains in an ecosystem. **Food webs** are used to illustrate the interaction between food chains. Sometimes their paths cross directly, and sometimes they do not. A food web helps you understand that the presence of any one species nearly always affects others. Look at the food web in Figure 3.7. What impact would a decrease in the rabbit population have on the other organisms in the community? A reduction in the rabbit population would affect both the snake and the owl, since both of these animals are predators of the rabbit.

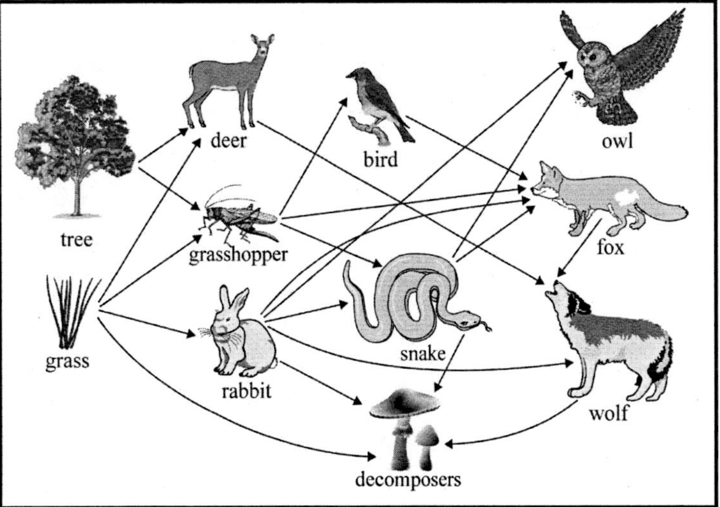

Figure 3.7 A Food Web

How about the grass? You can see by looking at the food web that the grass and trees (producers) feed everything to the right of them, either directly or indirectly. What if the grass were gone? Well, it has happened before. Decades of overproduction and inappropriate farming techniques stripped the fertile soil from the Great Plains in the early 1930s. This was an ecological disaster. Not only did the area become the Dust Bowl, there was also no food for primary consumers to eat. The consumers (both people and animals) were forced to move, leaving the barren wasteland behind.

Figure 3.8 The Dust Bowl

A **trophic level** is the position occupied by an organism in a food chain. Organisms that share a trophic level get their energy from the same source. Producers are found at the base of the energy pyramid and comprise the first trophic level of the food chain. Producers capture energy as sunlight and convert it into usable forms. Above them are the primary consumers that make up the second trophic level. Above the primary consumers are the secondary consumers that occupy the third trophic level. Finally, there are the tertiary consumers at the top trophic level. The tertiary consumers are the so-called "top" of the food chain. They are generally omnivores, like humans, or carnivores, like wolves. Different ecosystems will have different tertiary consumers. Figure 3.9 shows these relationships.

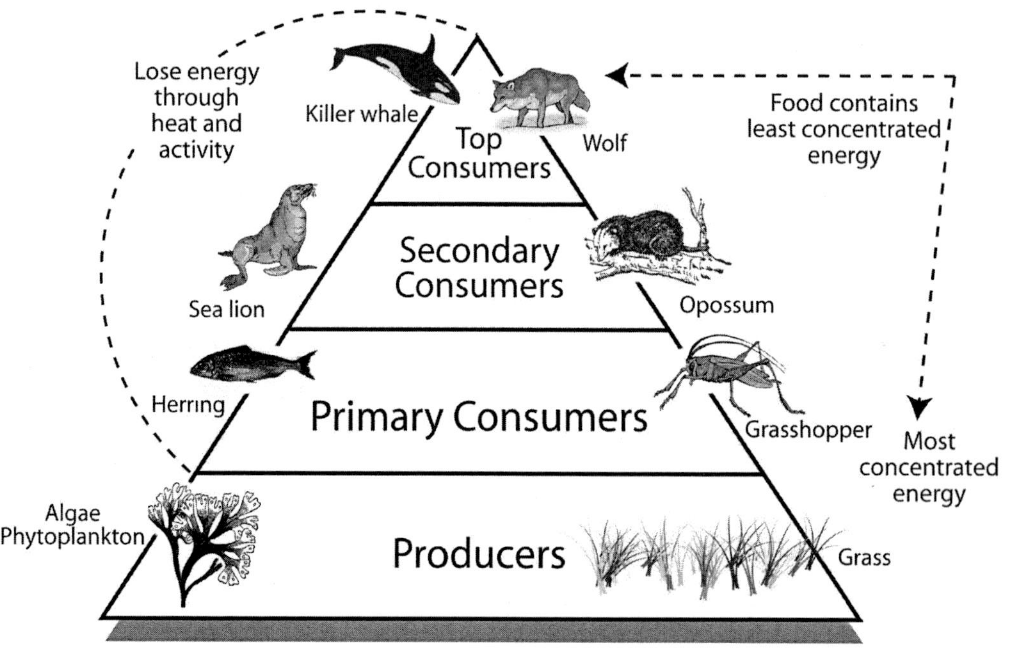

Figure 3.9 Energy Pyramid

It is important to say that food chains, food webs and energy pyramids are used to demonstrate to students how energy flows through ecosystems. They represent a single snapshot in time of an ecosystem. Realistically, the relationships between organisms are very complex. Animals often find food from many different sources. Even top consumers like wolves have been known to scavenge food, thus making a carnivore (and top consumer) into a decomposer. In another example, think of a grizzly bear — at times it is considered a primary consumer when it eats berries and grasses. Other times it is a secondary consumer when it eats insects, fish or small mammals. It can also be considered a decomposer when it scavenges meat. In general, food chains and food webs are good tools to help us remember an organism's place in the ecosystem.

Section Review 2: Food Chains and Food Webs

A. Define the following terms.

food chain	decomposer	carnivore	trophic level	omnivore
producer	herbivore	food web	consumer	saprophytes

B. Choose the best answer.

1 Identify two organisms below that share the same trophic level.

A elephants and lions
C chipmunks and squirrels
B cheetahs and giraffes
D wolves and sparrows

2. The owl is a nocturnal hunter of small mammals, insects and other birds. An owl is an example of a(n)

A producer. B omnivore. C carnivore. D decomposer.

3. Which food would an herbivore ***always*** avoid?

A worms B clover C pine nuts D grass

4. Emperor penguins feed mostly on crustaceans, such as krill. They are prey to orca whales and leopard seals. What ecological role does the emperor penguin play? (HINT: Krill are zooplankton, tiny sea organisms that feed on plankton.)

A It is a producer.
C It is a secondary consumer.
B It is a primary consumer.
D It is a top consumer.

5. Which trophic level contains the ***least*** amount of biomass?

A producer
C secondary consumer
B primary consumer
D top consumer

6. Snakes are ***always*** carnivores. At what trophic level do snakes belong?

A producer
C secondary consumer
B primary consumer
D decomposer

C. Fill in the blanks.

1 Animals that eat both plants and other animals are called _________________.

2 Organisms that obtain food from dead organisms or waste material are called _______________.

PLANT ADAPTATIONS

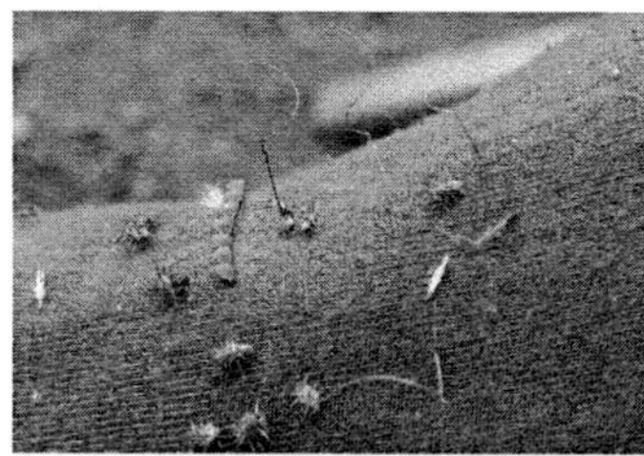
Figure 3.10 Animal Seed Dispersal

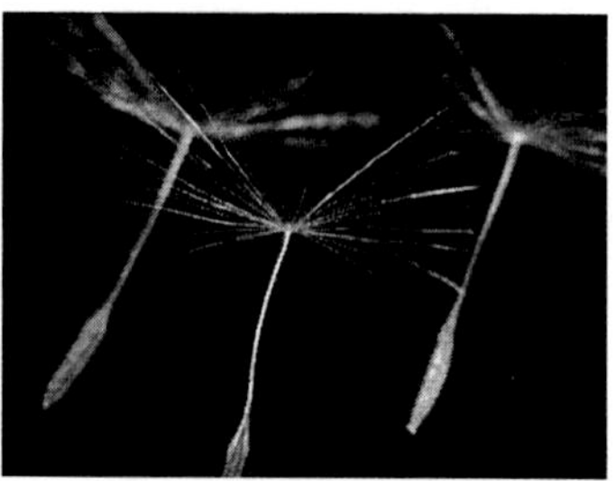
Figure 3.11 Wind Seed Dispersal

Plants cannot flee from predators, but they do have spines, thorns and leathery leaves to discourage herbivores from consuming them. Some plants manufacture chemicals that are poisonous or have a foul odor to keep animals away. Milkweed, tobacco and peyote cactus are three such plants. The Venus flytrap plants have a particular way of gathering food. They capture insects within their modified leaves. Some plants have developed uncommon methods of spreading seeds. Some plants have seeds with spines that attach to animal fur. Some have seeds with "wings" or parachutes used to harness the wind, and some seeds have watertight, buoyant outer shells used to travel in the water.

Mechanical stresses, such as wind, rain and animals, have an effect on the growth of plants. Indoor plants will grow taller than outdoor plants of the same species because they are protected from the weather. Adaptations to mechanical stress include shorter, thicker stems, which helps outdoor plants withstand the elements and increases their survival chances, even if the plants' overall growth is inhibited.

Challenge Question

What are some plant adaptations in your area? Jot a quick list here:

BEHAVIORAL PLANT ADAPTATIONS

Response to internal and external stimuli by an organism is called **behavior**. Plants respond to stimuli in a variety of ways to increase their chances for survival.

Recall that **tropisms** are the growth of a plant in response to a stimulus. **Positive tropisms** are toward the stimulus and **negative tropisms** are away from the stimulus.

Figure 3.12 Plant Phototropism

Geotropism is a plant's response to gravity. Roots are positively geotropic. They grow toward the Earth in response to gravity. Stems and leaves are negatively geotropic. Plants also respond to touch. This is called **thigmotropism.** Thigmotropism is the least common type of plant tropism. Climbing plants, like kudzu, honeysuckle or beans use thigmotropism. They generally have weak stems and will wrap around another plant, wall, fence or other structure for support. The tentacles of climbing plants respond to the touch of something else and coil around the object, sometimes in a matter of hours.

Other Plant Behaviors

Figure 3.13 Venus Fly Trap

Nastic movements are the responses of plants to stimulus regardless of direction. Examples include flowers opening and closing in response to light, or mimosa leaves curling up when touched by an object or blown by wind. Carnivorous plants, like the Venus flytrap, will close in response to something touching the little hairlike structures inside their leaves, which helps them obtain food.

Plants also follow circadian rhythms. **Circadian rhythms** are behavior cycles that follow roughly 24-hour patterns of activity. Some plants fold their leaves and flowers during the night and open them during the day, to preserve water. For many plants, light stimulates growth hormones. Other plants have adapted to secreting perfumes and nectars at times when their pollinators are active, increasing chances of fertilization.

Plants can send communication signals to predators and other plants. They have the ability to secrete foul-tasting substances, so herbivores will avoid eating them. Even leaves damaged by an herbivore can secrete the chemical warning to other plants. The other plants then produce the chemical substance. The plants only secrete the protective substance when needed.

Flower blooming follows a photoperiodic trend. **Photoperiodism** is the response of plant processes to the amount of daylight. Photoperiodism explains why plants bloom in different seasons. The amount of daylight in fall and winter is less than the amount of daylight in spring and summer.

Animal Adaptations

Physical Animal Adaptations

Physical adaptations help animals survive and flourish in their environment. Many species living in cold climates grow thick fur during the winter and shed it during the summer months. Animals that live in cold climates typically have short extremities (limbs, ears, tails). Shorter extremities reduce heat loss from an animal's body. Animals that live in dry deserts like camels, giraffes and African deer can survive for long periods without water. Some desert frogs have special adaptations that allow them to absorb water directly from the air; some frogs can change color to a pale white to reflect heat during the hottest part of the day. Animals that live in hot climates typically have large, long extremities, allowing them to cool off easily. Animals living in cooler climates have short, compact extremities that retain body heat. In one example, African elephants have large ears, while their Asiatic cousins have much smaller ears.

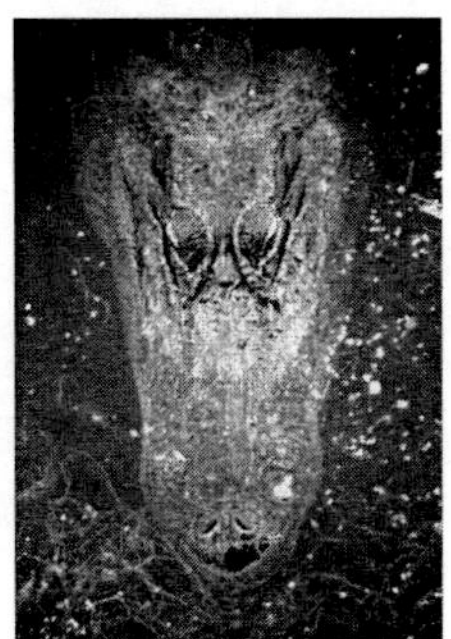

Figure 3.14 Alligator

Animals also adapt in response to predator/prey interactions. Porcupines and spiny anteaters grow sharp quills for protection against predators. Turtles retreat inside a bony shell for protection. Some animals produce venom or poison for hunting or protection. Armadillos have armorlike skin that protects the animal when it flees from predators into thorny patches. The American alligator has eyes and nostrils located on the top of its head, allowing the body of the animal to remain hidden beneath water. Cottonmouth snakes and other pit vipers have special heat-sensing organs, located on the front of their heads.

Figure 3.15 Predator

Most land-dwelling predators have eyes located close together on the front of the head. This allows the predator to focus specifically on its prey. Most prey animals have eyes located far apart on either side of the head. This allows the prey to see more of the area around it, perhaps helping its escape.

Figure 3.16 Prey

BEHAVIORAL ANIMAL ADAPTATIONS

Animals use behavioral adaptations for survival and reproduction. **Territoriality** is a behavioral adaptation that insures adequate space and resources for reproduction. For example, male elephant seals battle for specific beach territories during the breeding season. When female seals arrive, they remain on the beach within the territory of a single male seal. Large, strong males typically have the largest territories and the most females. In this way, the male is assured to pass along his genes to a large portion of the next generation of elephant seals. Remember, not all elephant seal pups will survive to adulthood. Fathering many offspring is one way males ensure reproductive success. Many animals fight for territory, including fish, elephants, horses, deer, rams...really too many to list!

Figure 3.17 Territorial Elephant Seals

Many animals have adapted behaviors that protect them from predators. Birds flock together, fish school together and insects form swarms to increase their chances for survival.

Some animals give warning signals. Brightly colored frogs and insects warn others of their distasteful or poisonous nature. Rattlesnakes use loud rattling sounds to warn other organisms of their presence. Skunks use both distinguishing colors and foul odors.

Figure 3.18 Flock of Birds

Figure 3.19 Bird of Paradise

Courting behavior is a behavioral adaptation that helps to insure beneficial genes are passed along to offspring. Courting is an example of reproductive adaptation. Mates that can build the best nests, sing exuberant mating calls or have the brightest colors are healthy and strong and will likely produce the strongest offspring. The courting behavior of insects, birds, amphibians, mammals or fish can be complex visual or auditory displays. Lightning bugs display bright lights to attract mates. Birds build nests, do dances, sing songs or grow specialized feathers to attract mates. Birds of the genus *Paridisaeidae* are commonly called birds of paradise; males of this genus are renowned for their ornamental plumage. Frogs, alligators and whales are just a few examples of animals that call for mates with elaborate songs.

Some animals — usually males — engage in elaborate rituals to lure a mate. Many male mammals fight with other males, and some birds will decorate nests, perform dances or puff up colorful feathers. The females generally try to select the males with the best traits, and those genes are passed along to offspring. Pheromones also act to increase communication among organisms. Insects rely heavily on pheromones or chemical signals for communication. Ants use pheromones when foraging for food. This creates the chemical trail that they follow when collecting a food source. Surely you have observed ants marching in a line toward a tasty food item. This behavior is possible because of pheromones.

Animals, like plants, follow circadian rhythms. Some animals are active during the day. They are **diurnal** animals, like squirrels and blue jays. Animals active at night are **nocturnal**, like bats and raccoons.

Figure 3.20 Bear

Some animals hibernate, estivate or migrate to escape extremes in weather. **Hibernation** is a period of dormancy during cold months. When animals enter a period of **dormancy**, which is a period of biological rest or inactivity, food supplies are limited, and the animal lives off its fat stores. Metabolism, breathing and body temperature all drop to conserve energy. Growth and development also cease during the dormant period. Bears hibernate in winter. **Estivation** is dormancy in hot climates. Lungfish estivate. Other animals **migrate**, or move to new locations in response to weather changes to stay close to food sources. These animals, like geese, usually follow the same routes every year.

Some animals exhibit learning behaviors. **Learned behaviors** are a result of an animal's experiences. It is believed that only animals with complex nervous systems are capable of learning. Generally speaking, animals with long life spans and lots of parental care have many learned behaviors. For example, birds use a type of learning called **imprinting** where young birds rapidly associate certain sights and sounds with their parents. **Habituation**, another type of learning, occurs when an animal learns not to respond to repeated stimulus, like when dogs stop barking at familiar people. A third type of learning, called **classical conditioning**, occurs when an animal associates a stimulus with an uncommon response. For example, dogs learn to sit at the sound of a clicker.

Reproductive Adaptations

Courting is one example of a reproductive adaptation. Fish and amphibians use another behavioral adaptation called spawning. During this time, large groups of male and female animals cluster together and release their gametes into the water. Massive fish spawns can even be observed from airplanes. Reproducing in such large numbers helps guarantee some offspring will escape predators and survive.

In the plant kingdom, flowers are a common example of a reproductive adaptation. In dry areas, plants form hardy sex cells called pollen. To create a seed, pollen grains must enter the female part of the flower called the embryo.

Some animals use special structures for reproduction. Some desert-dwelling lizards have specialized male sex organs called hemipenes. They actually have two male sex organs! Male squid and octopuses have specialized appendages used solely for reproduction. Marsupials are an entire order of mammals that use a specialized pouch for reproduction. Indeed, reproductive adaptations are so diverse and plentiful that these are but a few examples.

Section Review 3: Adaptations and Behavior

A. Define the following terms.

diurnal	hibernation	adaptations	estivation	tropism
migrate	courting behavior	nocturnal	phototropism	behavior
territoriality	nastic movements	geotropism	photoperiodism	dormancy
	circadian rhythms	thigmotropism		hormone

B. Choose the best answer.

1 Which of the following is ***true*** about the connection between adaptations and homeostasis?

A Adaptations help animals efficiently maintain homeostasis.

B Adaptations make it more difficult to maintain homeostasis.

C Homeostasis discourages the development of adaptations.

D Homeostasis is independent of adaptation development.

2 How do microbes maintain homeostasis?

A by causing disease

B by recycling matter

C by causing fish kills

D by signaling hibernation

3 Some leaves contain compounds that when broken down become toxic to plants. Roots growing near these compounds change direction or stop growing altogether. How would you describe the behavior of these roots in response to this chemical?

A positive tropism

B negative tropism

C phototropic

D geotropic

4 Why is it beneficial for some animals to migrate?

A to avoid temperature extremes or food shortages

B to look for new predators

C to find better shelter

D to catch up with lost populations

C. Complete the following exercises.

1 Name two adaptations plants have developed to disperse their seeds.

2 Name two adaptations animals have developed to maintain internal temperatures.

3 Explain homeostasis in your own words.

ORGANIZATION OF ECOSYSTEMS

ECOSYSTEM

An **ecosystem** is the interdependence of plant and animal communities and the physical environment in which they live. The **biosphere** is the zone around the Earth that contains self-sustaining systems composed of biotic and abiotic factors. **Biotic** factors include all living things, such as birds, insects, trees and flowers. **Abiotic** factors are those components of the ecosystem that are not living, but are integral in determining the number and types of organisms that are present. Examples of abiotic factors include soil, water, temperature and amount of light. In order for an ecosystem to succeed, its biotic factors must obtain and store energy. In addition, the biotic and abiotic factors of the ecosystem must recycle water, oxygen, carbon and nitrogen.

Figure 3.21 Organization of Life

COMMUNITY

A **community** is a collection of the different biotic factors in a particular ecosystem. Communities include many different species of plants and animals that live in close proximity to one another. For example, in a marine ecosystem, a coral reef supports a large community of plants and animals. In this example, the community of fishes, shrimps, mammals, algae, sharks, corals, urchins, sea stars and clams all live together and interact with one another. A community might have very different types of plants and animals living in one area. The members of a community interrelate with each other. Deer grazing in a clearing in the forest may be alert to the activity or movement of birds that warn them of approaching danger. In turn, the birds may depend on the deer grazing in a clearing to disturb insects hiding in the grass, thus causing them to become visible.

Figure 3.22 A Forest Community

Each member of a community has its own **habitat**. A habitat is the dwelling place where an organism seeks food and shelter. A woodpecker lives in a hole in a tree. It eats the insects that live in the bark of the tree. A robin builds its nest and raises its young in the same tree. A mouse lives in a burrow at the base of the tree. An owl sleeps on a branch of the same tree. The tree supports a whole community of organisms and becomes their habitat. The habitat provides food and shelter for the members of the community. In turn, each species of the tree community has its own **niche**. A niche is the role that an organism plays in its community, such as what it eats and where it lives.

Population

A community of living things is composed of populations. **Populations** are made up of the individual species in a community. For example, in a forest community ecosystem, there are populations of various plant and animal species such as deer, squirrels, birds, insects and trees.

Figure 3.23 A Population of Deer

Species

A **species** is a group of similar organisms that can breed with one another to produce fertile offspring. Organisms of the same species share similar characteristics common to all organisms within the population. For example, all domestic cats can breed to produce kittens. Additionally, all domestic cats eat meat, can land on their feet, and have whiskers, retractable claws, canine teeth.

Some natural variation exists within all members of a species. Not all cats have whiskers of the same length or bodies of the same size. However, all domestic cats can breed to produce offspring.

Two organisms are part of a different species when they cannot breed and produce fertile offspring. A horse and a donkey can breed, but they produce a mule, which is almost always infertile. Therefore, a horse and a donkey are different species.

Figure 3.24 Cat Species

Activity

Read each situation and determine which component of ecosystems it illustrates. The first one is done for you. Terms to use: community, habitat, population, species, niche and biosphere.

1	In a freshwater ecosystem, plants like bald cypress, pitcher plants and water lilies can be found. Animals like black bear, river otter, crane and white-tailed deer may also be present.	Community
2	A large school of Atlantic cod	
3	A polar bear resides in the area of the Arctic that is frozen by annual ice packs.	
4	The grey bat is an endangered bat found in Georgia.	
5	The spotted hyena (*Crocuta crocuta)* is a wild animal living in the South African savanna. This animal sometimes scavenges kills from other large carnivores. The hyena lives in a clan with other hyenas and has been known to hunt and kill large herbivores like wildebeest, zebra, Thomson's gazelle, Grant's gazelle, antelope and buffalo.	
6	In a terrestrial ecosystem, plants like saguaro, cardon, ironwood and ocotillo abound. Animals like coyotes, mountain lions, rodents, lizards, iguanas, Gila monsters and owls can be found living together.	
7	A large geographic area contains many self-sustaining systems.	
8	In a forest ecosystem, squirrels live an arboreal lifestyle and eat plant matter, insects, eggs and fungi. In the fall, they bury stores of tree seeds and nuts in the ground, helping to propagate many tree species.	
9	Annually millions of wildebeests migrate 500 to 1,000 miles.	
10	The three-toed sloth is a mammal that is usually found in the middle or upper layer of the rainforest canopy.	

Section Review 4: Organization of Ecosystems

A. Define the following terms.

ecosystem	abiotic	niche
biosphere	community	population
biotic	habitat	

B. Choose the best answer.

1 The area in which certain types of plants or animals can be found living in close proximity to each other is called a

A habitat. B community. C niche. D kingdom.

2. A British ecologist stated the importance of realizing an organism's role in the ecosystem as follows: "When an ecologist sees a badger, they should include in their thoughts some definitive idea of the animal's place in the community to which it belongs." What does this statement describe?

A an animal's habitat
C an animal's community
B an animal's niche
D an animal's ecosystem

3. The giant noctule bat (*Nyctalus lasiopterus*) preys mainly upon insects during the summer months, and on migrating songbirds during the autumn and spring. It attacks the birds at night from several hundred meters in the air. During the day, the bat roosts in trees. What do these sentences describe?

A community B habitat C biome D niche

4. Nitrogen, oxygen and carbon dioxide are among the ***most*** biologically important atmospheric gases. What are these called?

A abiotic factors
C biospheric factors
B biotic factors
D habitat factors

5. A hinny is the offspring of a male horse and a female donkey. Like mules, hinnies are ***almost always*** sterile (unable to breed). This confirms that

A a mule and a donkey are different species.

B a mule and a hinny are different species.

C a horse and a donkey are different species.

D a horse and a hinny are different species.

C. Complete the following exercise.

1 Name four abiotic conditions that might determine the kind of ecosystem in an area.

Relationships among Organisms

Each organism in an ecosystem interrelates with the other members. These relationships fall into one of three categories: symbiosis, competition or predation.

Symbiosis

A **symbiotic relationship** is a long-term association between two members of a community in which one or both parties benefit. There are three types of symbiotic relationships: commensalism, mutualism and parasitism.

- **Commensalism** is a symbiotic relationship in which one member benefits, and the other is unaffected. Hermit crabs that live in snail shells are one example of this type of relationship. Some argue that one member in most commensalistic relationships is harmed in some way by the relationship. For example, orchids are a type of epiphytic plant that grows on top of the branches of rainforest trees. Having a *few* orchids in its branches does not harm the tree. However, the accumulation of many epiphytic plants could cause the tree to lose limbs, sunlight or nutrients. For this reason, true commensalistic relationships are rare in nature.

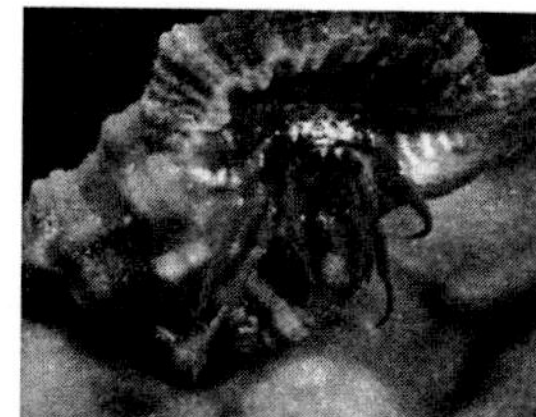

Figure 3.25
A Commensalistic Relationship

- **Mutualism** is a symbiotic relationship that is beneficial to both organisms. In South America, the tree *Acacia cornigera* and a species of ant (*Pseudomyrmex ferruginea*) are one example of mutualism. The acacia tree provides a home for the ants by growing specialized hollow thorns. The tree also provides food for the ants in the form of a protein-lipid and a carbohydrate-rich nectar from structures on the leaf stalk called nodules. The ants in turn protect the tree from predators by biting or stinging them and from other plant competitors by killing all plant life that comes into contact with the branches. There are many examples of mutualism in nature including: remoras and sharks or clown fish and sea anemones.

- **Parasitism** is a symbiotic relationship that benefits one organism (the parasite), but harms the other (the host). For example, heartworms in dogs (and humans) are parasites. The heartworm benefits by getting its nutrition from the bloodstream of its animal host. The host, however, is harmed because blood flow is restricted and nutrients are lost to the parasite.

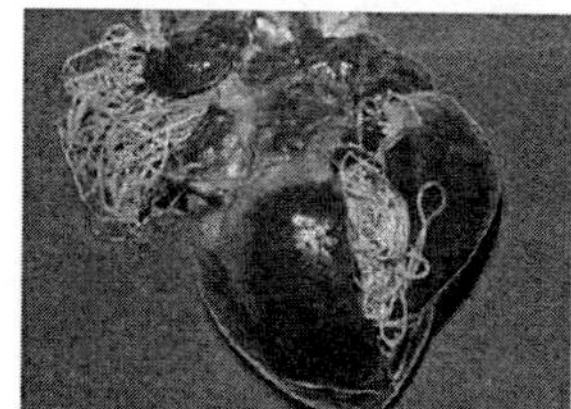

Figure 3.26 A Parasitic Relationship

Competition

When two or more organisms seek the same resource that is in limited supply, they **compete** with each other. A **resource** could be food, water, light or space. Competition can be between members of the same species or between members of different species. Competition is one of the main mechanisms behind natural selection.

Predation

Predators and prey help maintain an ecological balance within their ecosystem. This balance benefits the community as a whole, but can be helpful or harmful to the members that make up the community, depending upon whether they are the predator or the prey. A **predator** is an organism that feeds on other living things.

The organism it feeds on is the **prey**. For instance, wild dogs will hunt down and kill zebras, separating out weak and sick animals from the herd. The predator/prey relationship is the way energy passes up the food chain of the ecosystem, and is the main driving force behind natural selection.

Figure 3.27 Predator-Prey Relationship Between Wild Dogs and Zebras

Alternative Assessment Activity

Make a foldable about different symbiotic relationships of organisms. A foldable is a flip chart used to review topics. Begin by folding a blank sheet of paper in half lengthwise. The cut two slits along top, outside page only (see the illustration provided). You will now have three outer flaps. On each outer flap, write the term you are studying. For example, on the first flap write the word commensalism. On the second flap, write mutualism and on the third flap write parasitism. Then on the inside flap, write a description or definition of the term along with your own example.

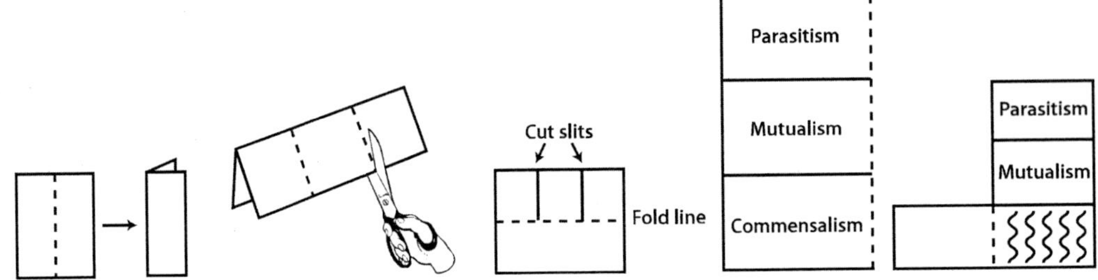

Use your foldable to quickly review these terms.

Section Review 5: Relationships among Organisms

A. Define the following terms.

symbiotic relationship	parasite	predation
commensalism	host	predator
mutualism	competition	prey
parasitism		

B. Choose the best answer.

1 The relationship between two members of a community in which one member harms another by its presence is

A parasitism.
B commensalism.
C mutualism.
D dependency.

2. A bee goes from flower to flower, gathering nectar. At each stop, the furry body and legs accumulate pollen from the flower, which the bee transfers as it moves. The flower needs pollen to reproduce, and the bee needs nectar to eat. What kind of relationship is this?

A parasitism
B mutualism
C commensalism
D predation

3. A mother cuckoo lays her egg in the nest of a warbler, then flies away. The warbler raises the baby cuckoo along with her own babies. The cuckoo baby grows quickly, becoming massive compared to the warbler babies. At some point, the baby cuckoo pushes the warbler babies out of the nest in order to make more room for itself. What does this scenario describe?

A parasitism
B predation
C mutualism
D abiotic

4. Which of the following is ***not true*** regarding predation?

A Predation helps maintain an ecological balance.

B Predators keep the numbers of prey animals under control.

C Predators choose the sick and weak prey because they are easier to catch.

D Predators choose the sick and weak prey because they are trying to maintain ecological balance.

Population Dynamics

A population is a group of organisms of the same species living in the same geographic area. Important characteristics of populations include the growth rate, density and distribution of a population. The study of these characteristics is called population dynamics.

Growth

The **growth rate** of a population is the change in population size per unit time. Growth rates are typically reported as the increase in the number of organisms per unit time per number of organisms present. The size of a population depends on the number of organisms entering and exiting it. Organisms can enter the population through birth or immigration. Organisms can leave the population by death or emigration. Immigration occurs when organisms move into a population. Emigration occurs when organisms move out of a population. If a population has more births than deaths and immigration and emigration rates are equal, then the population will grow. Ecologists observe the growth rate of a population over a number of hours, years or decades. It can be zero, positive or negative. Growth rate graphs often plot the number of individuals against time.

A population will grow exponentially if the birth and death rates are constant and the birth rate is greater than the death rate. **Exponential growth** occurs when the population growth starts out slowly and then increases rapidly as the number of reproducing individuals increase. Exponential growth is also sometimes called a **J-shaped curve**. In most cases, the population cannot continue to grow exponentially without reaching some environmental limit such as lack of nutrients, energy, living space and other resources. These environmental limits will cause the population size to stabilize, which we will discuss shortly.

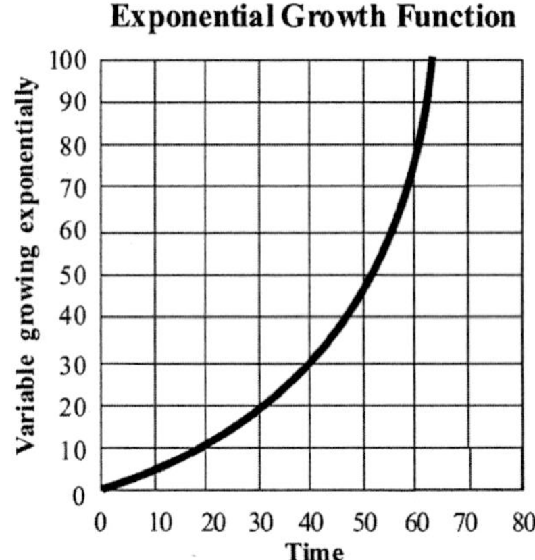

Figure 3.28 I-Shaped Curve Showing Population Growth vs. Time

Carrying Capacity

As the population uses up available resources, the overall growth of the population will slow or stop. Population growth will slow or decrease when the birth rate decreases or the death rate increases. Eventually, the number of births will equal the number of deaths. The **carrying capacity** is the number of individuals the environment can support in a given area. A thriving population will fluctuate around the carrying capacity. When the population size exceeds the carrying capacity, the number of births will *decrease* and the number of

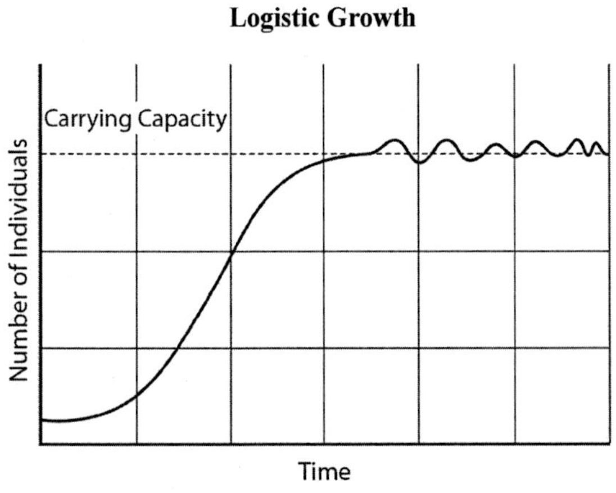

Figure 3.29 Carrying Capacity Graph

deaths will *increase*, thus bringing the population back down to the carrying capacity. This type of growth curve is known as **logistic growth**. Logistic growth is sometimes called an **S-shaped curve** because it levels out at a certain point.

Figure 3.30 Frogs

For example, let's say a specific pond has a carrying capacity of 40 frogs. If more than 40 frogs are in the pond, then food and space become limited. Some frogs will need to move to another pond, or members of the population will die. If fewer than 40 frogs are present in the pond, more frogs may move into the pond or more offspring will survive.

A decrease in environmental quality will decrease the carrying capacity of that environment. In the example above, if the pond becomes polluted, it will likely not be able to support 40 frogs. Its carrying capacity will be reduced to a population of fewer than 40.

An increase in the environmental quality will increase the environmental carrying capacity. For instance, if the pond is cleared of some or all of its pollution, it will again be able to support 40 or more frogs.

Regulation of Population Size

Availability of resources is not the only factor that limits population growth. A **limiting factor** is anything in a population that restricts the population size. Remember that resources in an ecosystem are limited, and the availability of matter, space and energy is finite. There are two main categories of limiting factors: **density-dependent factors** and **density-independent factors**. Density-independent factors are limiting no matter the size of the population, and include unusual weather, natural disasters and seasonal cycles. Density-dependent factors are *phenomena*, such as competition, disease and predation, which only become limiting when a population in a given area reaches a certain size. Density-dependent factors usually only affect large, dense populations. Disease is a major factor in many populations. It can quickly disrupt ecosystem balance. Dutch elm disease is one example. When this disease was introduced in the United States, the population of elm trees crashed.

Succession

Over time, an ecosystem goes through a series of changes known as **ecological succession**. Succession occurs when one community slowly replaces another as the environment changes. There are two types of succession: primary succession and secondary succession.

Primary succession occurs in areas that are barren of life because of a complete lack of soil. Examples are new volcanic islands and areas of lava flows such as those on the islands of Hawaii. Areas of rock left behind by retreating glaciers are another site for primary succession. In these areas, there is a natural reintroduction of progressively more complex organisms. Usually, lichens are the first organisms to begin to grow in the barren area. Lichens hold onto moisture and help to erode rock into soil components.

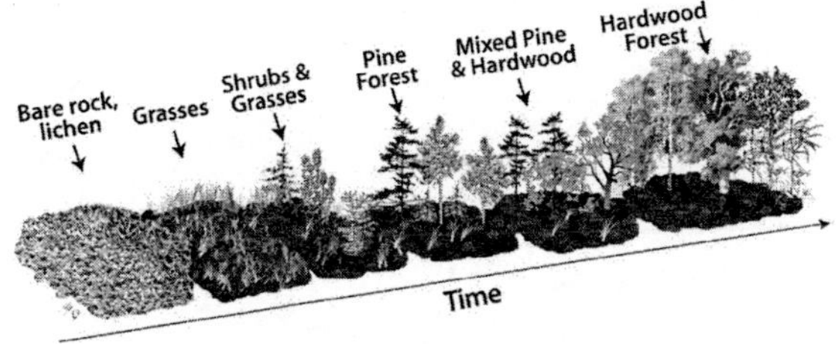

Figure 3.31 Primary Succession

The second group of organisms to move into an area, bacteria, protists, mosses and fungi, continue the erosion process. The chemical processes of these pioneer organisms, coupled with the decomposing matter created when they die, begins to create substantial amounts of soil. Once there is a sufficient number of organisms to support them, the insects and other arthropods inhabit the area. Grasses, herbs and weeds begin to grow once there is a sufficient amount of soil; eventually, trees and shrubs can be supported by the newly formed soil.

In habitats where the community of living things has been partially or completely destroyed, **secondary succession** occurs. In these areas, soil and seeds are already present. For example, at one time prairie grasslands were cleared and crops planted. When those farmlands were abandoned, they once again became inhabited by the native plants. Trees grew where there were once roads. Animals returned to the area and reclaimed their natural living spaces. Eventually, there was very little evidence that farms ever existed in those parts of the prairies.

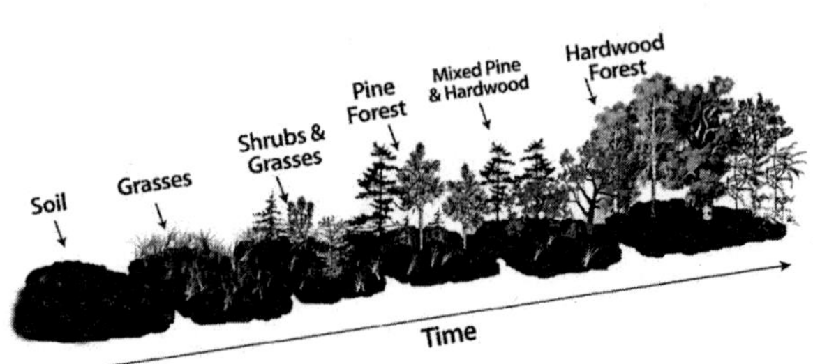

Figure 3.32 Secondary Succession

Ecosystem Stability

Earlier we discussed the idea of carrying capacity. Remember that populations do not exist by themselves. They are subjected to pressures within the ecosystem. For example, if persistent drought reduced the amount of vegetation in an area, it would impact all the other populations. Limited vegetation would reduce the herbivores in the area. This in turn, reduces the population of predators that depend on the herbivores. Recall that populations are regulated by food availability, space, disease, weather, and other populations. Although individual population sizes change, the ecosystem as a whole remains relatively stable. This is because ecosystems have a diversity of species. A healthy forest is not made up of only one tree species. A multitude of tree and shrub species all work together to create a functioning forest ecosystem. During some years, oak trees may dominate, while other years maple trees may be more numerous. Over hundreds or thousands of years, the forest system remains relatively stable. Ultimately, ecosystems maintain stability through diversity. Having a varied number and type of organisms in an area helps to guarantee the success of the system as a whole.

Section Review 6: Population Dynamics

A. Define the following terms.

growth rate	carrying capacity	density-independent factor
immigration	logistic growth	ecological succession
emigration	limiting factor	primary succession
exponential growth	density-dependent factor	secondary succession

B. Choose the best answer.

1 A density-dependent factor

A limits a population in a given area regardless of size.

B limits the population when the population reaches a certain size.

C may include weather or a natural disaster.

D often affects small, sparse populations.

2. How do ecosystems maintain stability?

A through succession

B through diversity

C through carrying capacity

D through exponential growth

3. A population will tend to grow if

A it is at its carrying capacity.

B it has a random population distribution.

C the number of births exceeds the number of deaths.

D the number of deaths exceeds the number of births.

4. An active volcano under the ocean erupts, and the buildup of cooled lava eventually forms a new island. What type of succession will immediately occur on the newly formed island?

A primary succession

B secondary succession

C conservation

D adaptation

C. Answer the following questions.

1 How is the carrying capacity of a population determined?

2. Why do you think it is important for a population to have limiting factors?

Chapter 3 Review

1 Lions are carnivores and are considered a _________________ in the energy cycle.

A primary consumer

B top consumer

C provider

D decomposer

2. Which organism is responsible for trapping atmospheric nitrogen?

A grass

B butterfly

C human

D bacteria

3. During the first part of the nitrogen cycle, bacteria converts the atmospheric nitrogen into what other compound?

A plant protein

B ammonia

C fertilizer

D nitrates

4. What are biotic factors?

A living factors

B lipids factors

C nonliving factors

D always unicellular

5. What are abiotic factors?

A decomposers

B living factors

C nonliving factors

D photosynthetic factors

6. What is the name for a place where a member of a community lives and finds food?

A pond

B biome

C habitat

D residence

7. How will unusual weather affect populations?

   A It will affect all populations regardless of size.

   B It will only affect small populations of organisms.

   C It will only affect large populations of organisms.

   D It will have no affect on populations.

8. Which of the following ***most likely*** would be a part of the first community on a newly formed volcanic island?

   A pine trees

   B oak trees

   C lichen

   D sea gulls

9. Red foxes are nocturnal and live in meadows and forest edges. They are predators to small mammals, amphibians and insects. The scraps that red foxes leave behind provide food for scavengers and decomposers. The preceding sentences describe the red foxes'

   A community.

   B prey.

   C niche.

   D food web.

10. Man-of-war fish cluster around the venomous tentacles of jellyfish to escape larger predators. The presence of the man-of-war fish does not harm or benefit the jellyfish. What is this type of relationship called?

    A parasitism

    B commensalism

    C succession

    D mutualism

Ecosystems

For questions 11 – 13, examine the diagram below.

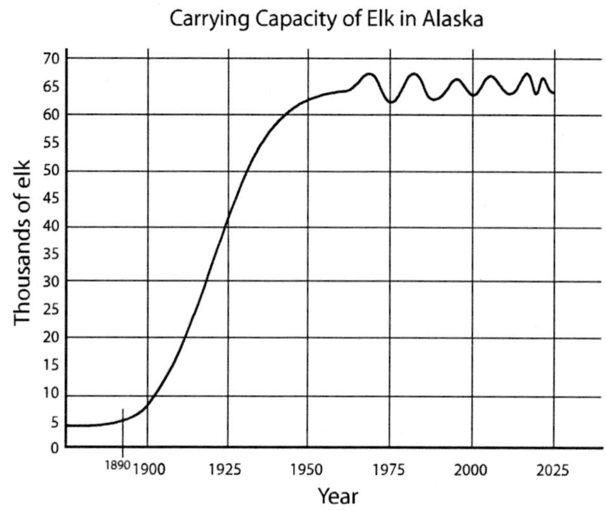

11. What is the carrying capacity for elk in this environment?

    A 65

    B 6,500

    C 75,000

    D 65,000

12. If a large oil company enters this environment and begins drilling for oil, building structures and polluting the land, what will probably happen to the carrying capacity of the elk?

    A It will be more than 65,000.

    B It will be less than 65,000.

    C Nothing; it will remain the same.

    D The elk will all leave and move into a new environment.

13. The United States government established this ecosystem as a native tribal reserve. Hunting is not permitted on native lands by anyone other than the native peoples. Based on the graph above, at what time was this ecosystem ***likely*** to have become a protected land?

    A 1962

    B 1950

    C 1925

    D 1890

14. A symbiotic relationship means

    A the energy cycle is not involved.

    B no one benefits.

    C the solar system is involved.

    D one or both parties benefit.

15. Which type of organism is responsible for converting atmospheric carbon into living systems?

    A humans

    B trees

    C mushrooms

    D bacteria

Chapter 4 Human Actions

This chapter covers:
2.2.1, 2.2.2

HUMAN POPULATION GROWTH

Although we as humans live in houses, we still rely on the environment to provide life's necessities: food, fresh air, water and space. Over time, our actions impact the environment. It is up to us to determine if our impact is positive or negative. In the past decade, the human population reached 6 billion. The positive growth rate has continued unabated since the Industrial Revolution in the early 1900s. In fact, the human population is experiencing an explosive exponential growth curve. Advances in agriculture, medicine and sanitation have allowed the human population to grow so quickly. Along with this explosive growth comes increased pressure on the environment. Humans use lots of space for housing and food production. Examine Figure 4.1, showing the human population for the past 10,000 years. At some point in the future, the carrying capacity for the planet will be reached, and humans will experience a net global negative population growth of zero.

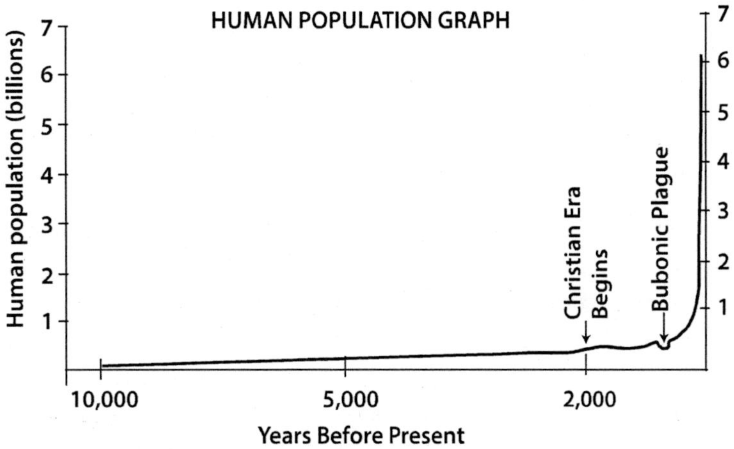

Figure 4.1 Human Population Graph

Because many world governments are concerned about population growth, they aggressively encourage population-control methods both voluntary (Western nations) and forced (China's "one child" policy) among their people. However, this promotion has led to a new, emerging problem, particularly in Europe at this point: underpopulation, especially of the youth. As a result of falling birthrates, an aging and retiring workforce has fewer youth and working-age populations to support the elderly in the population. European governments are finding it increasingly difficult to support the masses of elderly given the current workforce available.

Human Actions

The UN currently projects the world's human population to grow to 9.1 billion by 2050, but also shows the growth in population as slowing. Both government promotion of population control and voluntary changes on an individual basis have changed the level of growth, for better or worse.

Activity

Divide the class in half and debate the validity of government-sponsored population controls (like China's "one child" policy).

HUMANS AND THE ENVIRONMENT

There are many ways humans change ecosystems. Sometimes they change them for good. Sometimes they harm the ecosystem. Both ways change the environment for the organisms that live there. Environmental stewardship means taking good care of the Earth's ecosystems. Humans are responsible for managing the environment. We do this because we frequently cause the most harm. People also manage ecosystems because without a healthy environment we cannot survive. Ecosystems provide fresh air, water and food for *all* people.

Human actions can change the cycling of nutrients through the Earth's spheres. Humans have altered the nitrogen and phosphorous cycle through the use of chemicals. Fertilizers, detergents and pesticides all alter the natural cycling of nitrogen and phosphorus. When humans dam rivers, they slow water and increase its surface area which increases evaporation rates, effectively speeding up the water cycle.

POSITIVE HUMAN ACTIONS

A good steward positively affects the environment. When humans set aside land for national parks it is a positive action. The land returns to a natural state. It can now support many native plants and animals. Practicing sustainable agriculture is another positive action. This method of farming grows food without much environmental harm.

Figure 4.2 Historic Yellowstone Poster from 1938

Another way many people can protect the environment is through conservation and recycling. Many of the products you buy and use each day contain **natural resources**. Natural resources are materials people gather from the environment. They are used to make things for humans. There are many examples of natural resources in your classroom: paper, pencils, crayons, plastics and glass. Identify the natural resources used in the things found in your desk. Some resources are **renewable**. This means they can be replaced in your lifetime. They include wood, cotton or water. Other resources are **nonrenewable**. They cannot be replaced in your lifetime. They include coal, oil and natural gas.

Many natural resource materials can be **recycled**. When something is recycled, humans transform it into something new. Glass, plastic, aluminum and paper are often recycled. Many people also conserve, or use less of, these materials. Some natural resources cannot be recycled. Things like coal, oil or natural gas are natural resources used for fuel. They are collected from the Earth and consumed by people.

Negative Human Actions

As stated earlier, humans frequently cause the most environmental harm. People create pollution and litter and destroy habitats. All these actions have harmful environmental consequences.

Pollution

All types of pollution harm the environment and make people and animals sick. It can cause diseases or death. There are three main types of pollution: air pollution, soil pollution and water pollution. **Air pollution** is from people putting harmful things into the air. Cars, trains, buses and airplanes put lots of pollution into the air. Coal power plants and some factories also pollute the air. Air pollution can travel many miles and affect many different living things. Acid rain is one harmful consequence of air pollution. Acid rain is precipitation with unusually low pH. Acid rain can change the chemistry of soils and watersheds. This can harmfully change the habitat for many organisms. **Soil pollution** is caused by people putting harmful things into the soil. Landfills are one source of soil pollution. Throwing away chemicals can pollute the soil. Things like paint, motor oil, cleaners, prescription drugs and batteries should never be thrown away. Take these things to a proper collection center. **Water pollution** is from people putting harmful things into the water. Trash and chemicals often create pollution. Rain can carry soil pollutants into water. So, runoff from a rainstorm can also pollute water.

A **pesticide** is a chemical agent used to kill damaging or harmful organisms, usually animals. Examples of organisms controlled using pesticides include termites, fire ants, grasshoppers, snails, beetles, snakes and even other mammals. Pesticides are used on a commercial scale in agriculture. They alter the dominant species in an environment. Many homeowners also use pesticides to keep animals out of their homes and off their lawns.

Habitat Reduction

Figure 4.3 Logging Operation

Humans disrupt and use natural lands. This causes habitat reduction. Used by people, the habitat can no longer support other organisms. Clearing forests for neighborhoods, agriculture or logging disrupts woodlands. Plowing grassland for agriculture disrupts the prairie. Draining a swamp to build a neighborhood destroys the wetland. Agriculture, logging, mining and cities all destroy natural lands. Even suburbs displace native plants and animals.

Many cities and suburbs were once natural land. For example, New Orleans was once a swamp. It was drained to build the city. Draining the swamp displaced the plants and animals that once lived there. We call this **urbanization**. Urbanization decreases biodiversity of plants and animals in the ecosystem. In North Carolina, urbanization is particularly harmful in the Piedmont region. Development in this area causes damaging habitat destruction and water runoff.

Globally, agriculture is the main cause of habitat destruction. Humans need a lot of food to feed themselves and their livestock. Natural areas are converted to farms. This completely changes the ecosystem. Farms often grow only one type of crop. This destroys the variety of food found in a food web. Eventually, many organisms become endangered or threatened. In North Carolina, waste lagoons on hog farms are one troubling side effect of agriculture.

SPECIES INTERACTION

Figure 4.4 Brown Pelican

Endangered or threatened means very few organisms of that type are alive. The red-cockaded woodpecker, brown pelican, gopher tortoise and Louisiana black bear are some endangered species in the United States. Habitat destruction and pollution threaten these species. You will find these animals on the **endangered species** list. The endangered species list was created in 1963. An international group called the International Union for the Conservation of Nature (IUCN) maintains the list today. This list changes daily. Efforts to help the brown pelican have been mostly successful. In fact, it has been proposed that the brown pelican be taken off the list in the near future. There are many parks and rescue efforts to help America's other endangered species. Breeding programs or setting aside land are positive ways to help endangered species recover.

Figure 4.5 Jumping Carp

One aspect of modern life is the ability to travel great distances in a very short period of time. People can transverse the Atlantic or Pacific ocean in a matter of hours. When people travel they often bring organisms from one place to another, intentionally or mistakenly. Invasive, or **non-native species** are foreign organisms that enter ecosystems usually through human action. Invasive species can be plants, animals, fungi or microbes. There are thousands of examples of humans introducing non-native species into areas. When new species enter an area, they can displace native species through competition for food or space. Invasive species often have no natural predators in the new area, and their populations explode. Along with causing ecological harm, invasive species can also cause economic harm by damaging crops or livestock. Invasive species can reduce biodiversity by causing other more fragile species to become endangered or extinct. So, initially, it may seem that invasive species introduction increases biodiversity; however in the long run, it will decrease biodiversity. In this way, people can quickly cause long-term changes to ecosystems. Figure 4.5 shows jumping carp; these are invasive Asian carp that have become part of the Mississippi River system.

There are many examples of invasive species in several states. The Burmese python in Florida, kudzu in North Carolina and nutria in Louisiana are a few examples. Invasive species introduction is one way humans have most impacted the population of vertebrate animals.

Section Review 1: Humans and the Environment

A. Define the following terms.

pesticide	endangered	natural resource
pollution	invasive species	

B. Choose the best answer.

1 How has the human population changed in the last hundred years?

A It has grown.
B It has shrunk.
C It has remained constant.
D It has stopped.

2 Which of the following is a renewable resource?

A oil
B coal
C natural gas
D wood

3 What is a positive way humans impact endangered species?

A breeding program
B urbanization
C agricultural runoff
D algal bloom

4 Sloths live in the Amazon Rainforest. They live in the trees and eat mostly leaves. Many acres of the Amazon forest are being cleared for soybean farms. What will **likely** happen to the sloths?

A They will start eating soybeans.
B They will start walking on the ground.
C They will become endangered.
D They will begin hunting humans.

5 An invasive species

A helps create diversity.
B can take over an ecosystem.
C often helps humans in some way.
D generates the best organisms.

C. Answer the following questions.

1 Predict what is likely to happen to human population in the next 100 years. Support your predictions with evidence.

2. Describe one way you positively impact the environment.

3. Discuss why environmental concerns are important for everyone.

HUMAN ACTIONS AND GLOBAL WARMING

No doubt global warming, or global climate change, is something you've heard before. From print news to television media, global warming is something many people talk about today. What most people neglect to mention is that global warming is part of a *natural* Earth-warming process. This process is called the greenhouse effect.

The **greenhouse effect** traps solar heat within the Earth's atmosphere. Electromagnetic radiant energy provided by the Sun travels to the Earth through space. Some of this energy is trapped by bodies of water and land masses, and some of this energy is reflected back into the atmosphere. Some reflected heat is lost back into space but most is trapped by atmospheric gases. Gases like water vapor, carbon dioxide, methane and chlorofluorocarbons tend to trap more heat than other types of gases. Without these heat trapping processes, the average global temperature would be around –20 °C. However, like many other natural cycles, humans have altered this natural warming cycle as well.

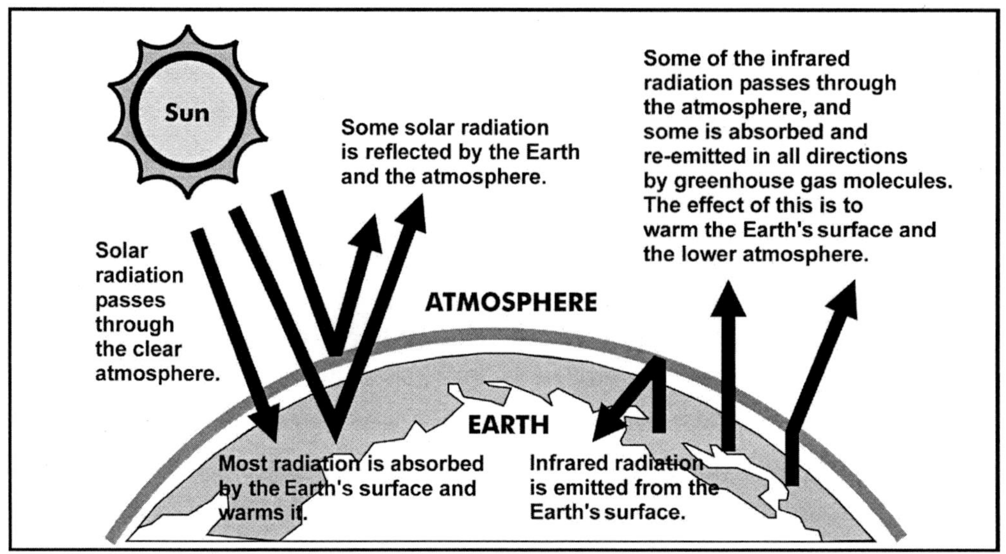

Figure 4.6 The Greenhouse Effect

Global warming is a term that describes the measured rise in the Earth's atmospheric and oceanic temperatures. The temperature rise is generally attributed to the increase in greenhouse gases in the atmosphere. Many argue that as humans burn fossil fuels, this increases the amount of greenhouse gases in the atmosphere. This is one of the many ways that human activity is changing our planet. The degree of human impact on climate and the best way to fix the problem are still hotly debated topics. As an active, inquiring member of the global community, you must examine the facts and decide where you stand.

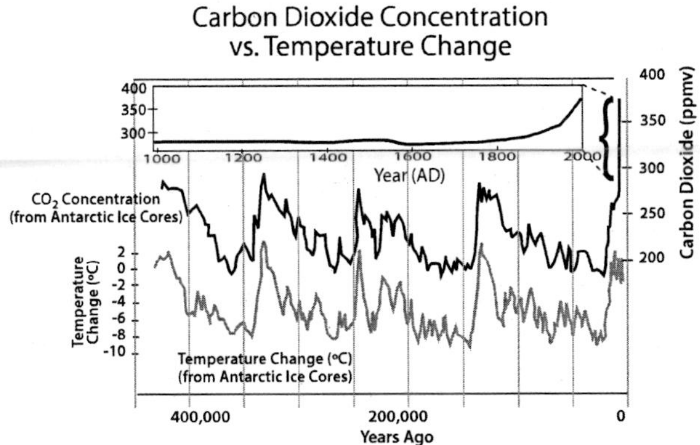

Figure 4.7 CO_2 Global Temp

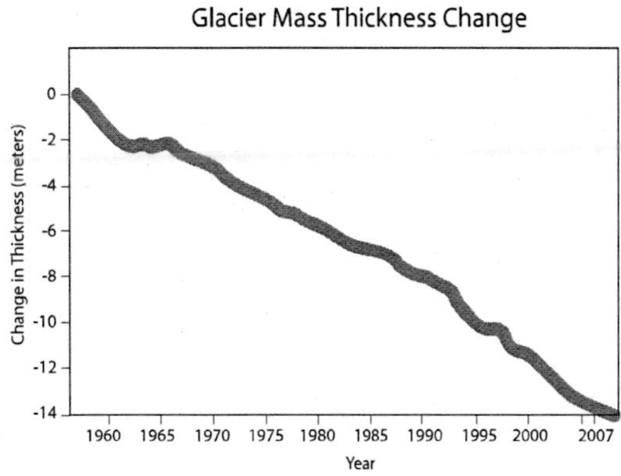

Figure 4.8 Average Global Glacier Mass Thickness

Figure 4.9 Arctic Ice Minimum 1979

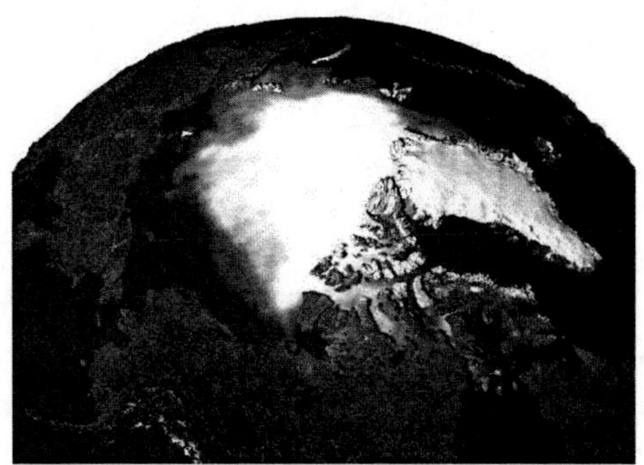

Figure 4.10 Arctic Ice Minimum 2005

Recall that inferring and concluding are two important aspects of scientific inquiry. Figures 4.7 – 4.10 can be interpreted in several different ways by several different people. Based on these observations and computer models, many scientific minds the world over agree that human-caused climate change is a real threat to the Earth's populations. Scientists expect to see increased temperatures, frequent intense storms and rising sea levels to impact human populations in the near future. The changes caused by global warming are predicted to cause the next global ice age. It is theorized that as polar ice melts, ocean currents will slow, causing a dramatic drop in global temperatures and thus beginning the next ice age. There is no doubt that burning fossil fuels adds carbon dioxide to the atmosphere. There is also little question that humans have altered this cycle. But to what extent these alterations have on the system remains to be seen. After examining the facts above and reading this section, what can you infer about global warming? The Earth will survive the next ice age — will we?

Human Consumption

Figure 4.11 Water Consumption

As human technology has advanced, so too has our ability to modify our environment to suit our needs. Humans need **natural resources** such as water, soil and air for survival. Humans dam rivers to prevent flooding and provide a consistent, reliable source of fresh water. Dams create lakes and increase the rate at which water evaporates, thereby speeding up the water cycle. Careless water consumption like running the water while you brush your teeth or having leaky pipes further increases the rate at which this precious natural resource is used.

Figure 4.12 Power Production

Along with the need for natural resources, humans also have a need for power, and lots of it. Humans use power to heat and cool homes, store and prepare foods and run electronic equipment. Humans can obtain power from natural resources like fossil fuels, water, sunlight and wind. The growing human population places increased demands on ecosystems to produce energy at ever increasing rates. In reality, humans are using up many natural resources much faster than they are recycled by ecosystems. Fossil fuels, a nonrenewable energy source, produce over 60% of the electrical power used in the United States. Careless power consumption, like leaving lights or television sets turned on when not in use, reduces the amount of energy available in the future and increases the amount of carbon released into the atmosphere (accelerating global warming trends). Non-renewable resources are finite and are in limited supply on Earth. Once used, fossil fuels cannot be replaced in a human lifetime, or even in several human lifetimes.

Activity

Think of activities that consume power and water in your home. Come up with ways to reduce your consumption of these resources, and begin using one or two each week. Record your experiences while attempting to implement your plan. After a month, write a paper or put on a play about your most meaningful moment during the "consumption reduction month."

Section Review 2: Human Actions and Global Warming

A. Define the following terms.

greenhouse effect global warming

B. Choose the best answer.

1 What is **not** an expected result of increased global temperatures?

A intense storms

B rising sea levels

C increased oxygen levels

D global ice age

2 Which factor listed below is ***not*** considered scientific evidence for global warming?

A data from retreating glaciers

B average increase in global temperatures

C increase in storm intensities

D increased rate of extinctions

3 Global warming

A is caused solely by humans.

B is not real.

C is a natural process.

D helps only tropical organisms.

4 How do humans ***most*** use power?

A in homes

B in cars

C in fossil fuels

D in airplanes

5. What is the ***most*** careless use of natural resources?

A watching TV at night

B using natural gas to heat your home in the winter

C having a nighttime security light at home

D leaving the water on while brushing your teeth

C. Answer the following question.

1 In your own words, explain global warming as a natural cycle.

CHAPTER 4 REVIEW

1 The greenhouse effect is responsible for which environmental condition listed below?

A clear-cutting of rainforests
B thermal pollution
C photosynthesis
D global warming

2. How has urbanization impacted species diversity?

A it has increased species diversity
B it has decreased species diversity
C it has remained unchanged
D it has increased then decreased

3. Which human action ***most*** impacts biodiversity of vertebrate animals?

A non-native species introduction
B global warming
C pesticide use
D strip mining

4. What is the proposed ultimate result of global warming?

A space exploration
B volcanic activity
C an ice age
D increased shark attacks

5. How can you ***most*** help to combat global warming?

A turn off lights when you leave a room
B keep the air conditioning on
C move to a cooler climate
D avoid eating fish

6. How has human agriculture impacted species diversity?

A It has increased species diversity.
B It has decreased species diversity.
C It has remained unchanged.
D It has increased then decreased.

7. Which of the following is used as evidence for global warming?

A chemicals in estuaries
B ocean mass extinctions
C algal blooms in freshwater
D global glacier mass

8. What is the ***most*** damaging human-caused environmental change?

A succession
B human consumption
C pesticide use
D convergence of tectonic plates

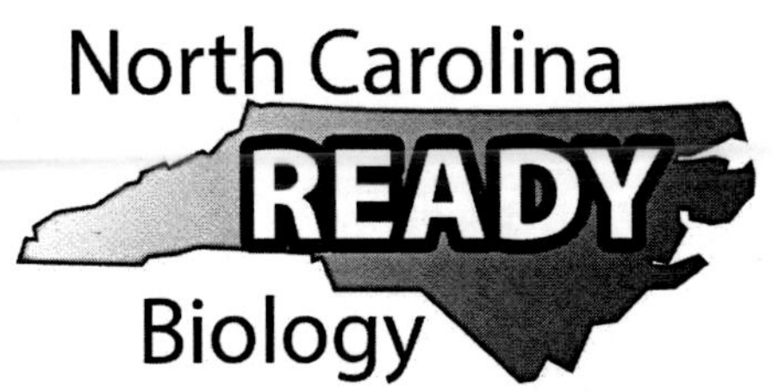

Chapter 5
DNA

This chapter covers:
3.1.1, 3.1.2, 3.1.3

The Structure of DNA and RNA

The genetic basis of life is a molecule called **DNA** or **deoxyribonucleic acid**. DNA is carried in the nucleus of all cells and performs two primary functions. First, it carries the code for all the genes of an organism, which in turn create the proteins that perform all the work of living. Second, the code of the DNA itself is the template for future generations. It transmits genetic information from one generation to the next. First, we will look at the structure of DNA and then its role in heredity.

DNA

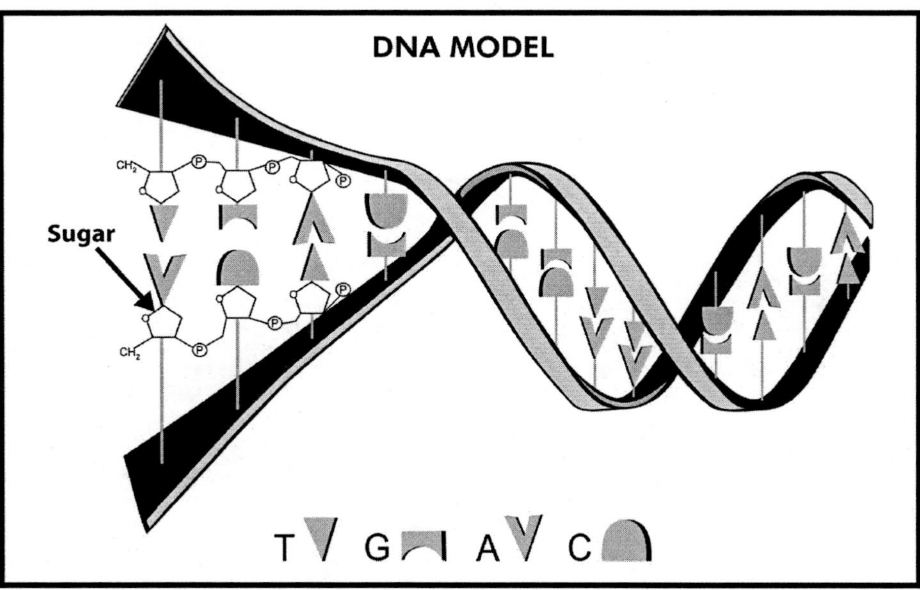

Figure 5.1 Model of DNA

DNA is a complex molecule with a double helix shape like a twisted ladder. Each side of the helix is composed of a long strand of **nucleotides** that are the building blocks of nucleic acids. Each nucleotide contains a phosphate group, the sugar **deoxyribose** and a nitrogenous **base**. There are four bases in DNA, and they form pairs. The bases are **adenine** (A), **thymine** (T), **guanine** (G) and **cytosine** (C). A and T always pair, and G and C always pair.

The A-T and G-C pairings are called **complementary pairs.** Each pair forms one of the rungs of the ladder. The two sides of the ladder are held together by hydrogen bonds between the nitrogen bases.

The DNA molecule carries the code for all the genes of the organism. **Genes** are pieces of the DNA molecule that code for specific proteins. The process of making genes into proteins is called **protein synthesis**. Proteins make the structures of an organism. Hair, nails and pigments (coloration) are just a few examples of proteins in the human body.

DNA is located in the nucleus of the cell. The assembly of proteins occurs outside of the nucleus, on the ribosome. So the manufacture of proteins involves three basic steps:

1 The DNA code of the gene segment must be copied in the nucleus of the cell.

2 The code must then be carried from the nucleus into the cytoplasm and finally to a ribosome.

3 The protein is then assembled from the code and released from the ribosome.

These steps are carried out by RNA, or ribonucleic acid. For more information on protein synthesis, view our downloadable animation at www.americanbookcompany.com/science or on TeacherTube at: http://www.teachertube.com/viewVideo.php?video_id=60707&title=Protein_Synthesis_Animation

RNA

RNA (ribonucleic acid) is a molecule used to translate the code from the DNA molecule into protein. It is similar to DNA, except it is single stranded. Its sugar is **ribose**. RNA, like DNA, has four nitrogenous bases. It shares adenine, guanine and cytosine but replaces thymine with **uracil** (U), so the bases A and U pair up instead of A and T. There are several types of RNA. Messenger, ribosomal and transfer RNA are all involved in creating proteins.

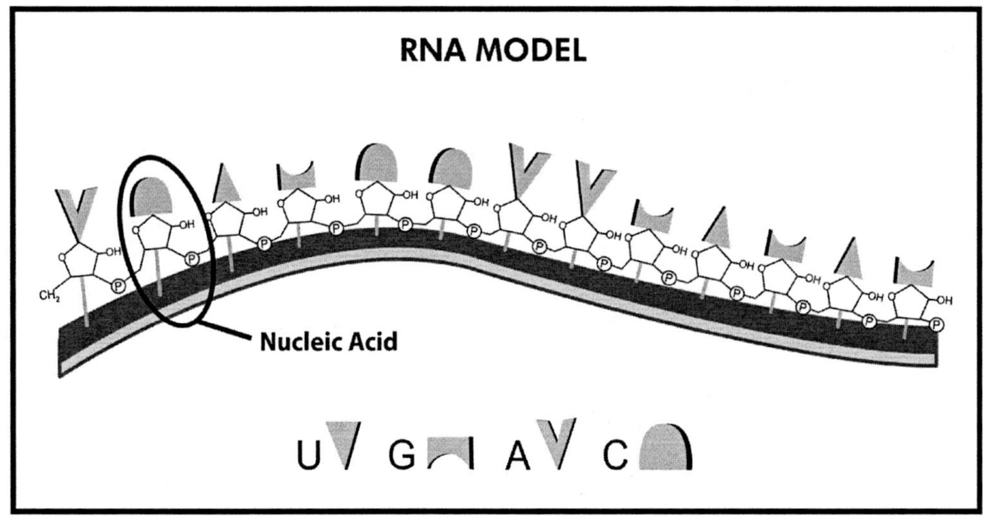

Figure 5.2 Model of RNA

Table 5.1 DNA and RNA Comparison

	Structure	**Sugar**	**Bases**
DNA	double stranded	deoxyribose	adenine **thymine** guanine cytosine
RNA	single stranded	ribose	adenine **uracil** guanine cytosine

Section Review 1: DNA and RNA

A. Define the following terms.

DNA	nucleotide	adenine	complementary pairs
gene	deoxyribose	cytosine	amino acid
RNA	base	uracil	thymine
guanine	ribosome	ribose	

B. Choose the best answer.

1 What forms the backbone of DNA?

A uracil B guanine C deoxyribose D ribose

2 What is an important part of both DNA and RNA?

A deoxyribose B ribose C uracil D nitrogen

3 Which DNA base is complementary to guanine?

A thymine B adenine C cytosine D uracil

4 Which DNA base is complementary to adenine?

A thymine B adenine C cytosine D uracil

5 Which force holds together the two sides of DNA?

A covalent bonds C ionic bonds

B hydrogen bonds D gravity

C. Answer the following questions.

1 Which sugars are found in DNA and RNA?

2 What are some examples of proteins?

Protein Synthesis[1]

TRANSCRIPTION

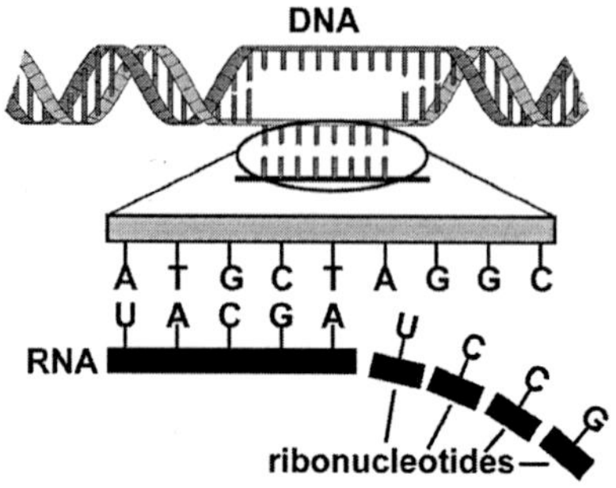

DNA

A T G C T A G G C

U A C G A U C C G

RNA

ribonucleotides

Figure 5.3 Transcription

There are many proteins within every cell. Proteins make up **enzymes** that help to carry out reactions within the cell. Proteins also compose **hormones**, which are chemical messengers that regulate some body functions. Proteins provide structure and act as energy sources. They transport other molecules and are part of our bodies' defenses against disease. In short, proteins are essential for survival because almost everything that happens in the cell involves proteins.

Transcription

The first step of protein synthesis is the manufacture of a specific kind of RNA called **messenger RNA (mRNA)**. This copying process is called **transcription**. Transcription begins when a region of the DNA double helix unwinds and separates, as shown in Figure 5.3. The separated segment is a gene, and it serves as a template for the soon-to-be-formed mRNA strand.

The mRNA strand is assembled from individual RNA nucleotides that are present in the nucleus. An enzyme called **RNA polymerase** picks up these unattached nucleotide bases and matches them to their complementary bases on the DNA template strand. This continues until the entire gene segment has been paired, and a complete mRNA strand has been formed. This mRNA strand has a sequence that is complementary to the original gene segment. At that point, the mRNA separates and leaves the nucleus, moving out into the cytoplasm to settle on the **ribosome**, an organelle composed of another kind of RNA, called **ribosomal RNA (rRNA)**. Here, on the surface of the ribosome, the process of translation begins.

Translation

Translation is the next step in protein synthesis where the mRNA molecule is decoded (translated) and a corresponding polypeptide is formed. (Remember that a polypeptide is made up of **amino acids**.) Let's look at the "language" of mRNA.

One way to think of a strand of mRNA is as a chain of nucleotides, as in:

AUGACAGAUUAG

While this is correct, another way of thinking of the chain is that it is divided into segments consisting of three nucleotides each, as in:

AUG ACA GAU UAG

The mRNA strand is not *actually* divided, but writing its code in this way emphasizes an important concept: the **codon**. The three-nucleotide codon has the specific function of corresponding to a particular amino acid. Here is how it works: The molecule of mRNA is bound to the surface of the ribosome at the first three-nucleotide segment, called the **start codon**. The cytoplasm in which they float contains, among other things, amino acids and a third kind of RNA — **transfer RNA (tRNA)**. Transfer RNA is a molecule of RNA that contains a three-part nucleotide segment called an **anticodon**, which is the exact complement of one mRNA codon. The anticodon corresponds exactly to one of the 20 kinds of amino acids. Once the tRNA binds the amino acid, it travels to the ribosome surface. There the three tRNA nucleotide bases (the anticodon) pair with their three complementary mRNA bases (the codon). The amino acid that is bound to the tRNA is then added to the growing polypeptide chain at the surface of the ribosome. The ribosome facilitates this process

1. To see a free downloadable animation of protein synthesis, go to http://www.americanbookcompany.com/science

by moving along the mRNA chain until it reaches a **stop codon**, a three-nucleotide segment that tells the ribosome that the translation process is complete. The ribosome then releases the newly formed polypeptide chain, which moves out into the cell as newly formed protein.

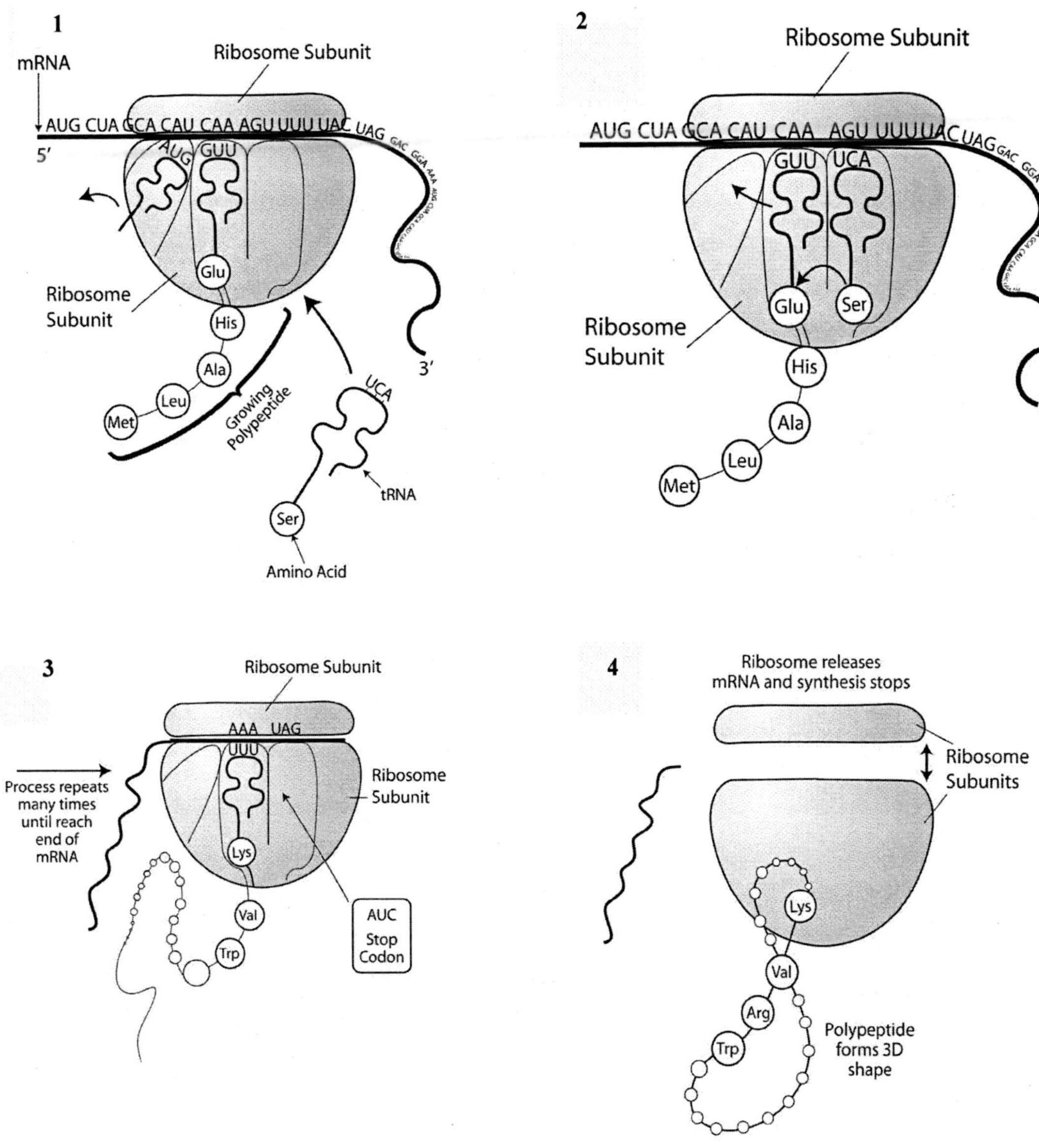

Figure 5.4 Translation

As you can tell, this is a complex process. It is the sequence of mRNA codons that determines the sequence of amino acid subunits. In particular, the chemical makeup of the amino acid's unique side chain ultimately determines their final bonding structure within the protein. Some side chains contain various elements like carbon, sulfur or nitrogen. They can be linear, branched or ring-shaped. Side chains are classified based on how they distribute their molecular charge: nonpolar, polar or charged. The interaction between the side chains of the individual amino acids eventually determines the 3-D structure of the protein. The structure of the protein establishes how it will function for the organism. It only makes sense that organisms would regulate the process of genetic expression in several ways. One of these ways is through transcription and translation. Without the proper amount of nucleotides and amino acids in the cytoplasm, genes cannot generate proteins.

USING A CODING DICTIONARY

The genetic code is the order of the bases that are part of the molecule of DNA. Once the genetic code is copied into a molecule of mRNA, the genetic code is read in groups of three bases called codons. Each codon specifies a particular amino acid to be added to the protein that will be built from the mRNA. We can use a **coding dictionary** to figure out which amino acids are coded for in a given mRNA chain.

A chart like the one in Figure 5.5 is called a coding dictionary; it shows which amino acid is coded for by a particular codon. There are twenty amino acids. If you have the codon GAU, you would begin by finding the first base (G) on the left-hand side of the chart. With a pencil, shade all the boxes to the right of G. Now, find the second base (A) on the top of the chart. Shade the boxes in the column underneath A. Note from the intersection of the two shaded areas that our amino acid can be either glutamic acid or aspartic acid. Finally, look up the third base (U) on the right-hand side of the chart to figure out which amino acid it is. If you look to the right, you'll see that GAU codes for aspartic acid. Once you have practiced, try reading the chart without actually shading it in with your pencil. Notice that some amino acids can be coded for by more than one codon.

First Position		Second Position: U	C	A	G	Third Position
	U	Phenylalanine	Serine	Tyrosine	Cysteine	U
		Phenylalanine	Serine	Tyrosine	Cysteine	C
		Leucine	Serine	Stop	Stop	A
		Leucine	Serine	Stop	Tryptophan	G
	C	Leucine	Proline	Histidine	Arginine	U
		Leucine	Proline	Histidine	Arginine	C
		Leucine	Proline	Glutamine	Arginine	A
		Leucine	Proline	Glutamine	Arginine	G
	A	Isoleucine	Threonine	Asparagine	Serine	U
		Isoleucine	Threonine	Asparagine	Serine	C
		Isoleucine	Threonine	Lysine	Arginine	A
		Methionine	Threonine	Lysine	Arginine	G
	G	Valine	Alanine	Aspartic acid	Glycine	U
		Valine	Alanine	Aspartic acid	Glycine	C
		Valine	Alanine	Glutamic acid	Glycine	A
		Valine	Alanine	Glutamic acid	Glycine	G

Figure 5.5 Coding Dictionary

Notice the codon AUG codes for methionine. AUG is a common "start" codon, which indicates the beginning of a gene. Now look at the codon UGA. This is one of three "stop" codons that indicates the end of a gene. Once a stop codon is reached, a protein is terminated. What are the other two stop codons?

Section Review 2: Protein Synthesis

A. Define the following terms.

DNA	nucleotide	ribose	amino acid
gene	deoxyribose	transcription	translation
RNA	base	messenger RNA (mRNA)	transfer RNA (tRNA)
anticodon	ribosome	ribosomal RNA (rRNA)	enzyme
protein synthesis	adenine	codon	hormone
polymerase	cytosine	complementary pairs	thymine
guanine	uracil	stop codon	start codon

B. Choose the best answer.

1 Which molecule starts the process of protein synthesis?

A mRNA B rRNA C tRNA D nucleotide

2 What are ribosomes made of?

A mRNA B rRNA C tRNA D protein

3 Proteins are made up of polypeptide chains. What are polypeptide chains composed of?

A mRNA B rRNA C tRNA D amino acids

4 What does transfer RNA (tRNA) carry?

A the mRNA to the ribosome

B the nucleotide bases to the mRNA

C an amino acid to the ribosome

D an amino acid to the cytoplasm

5 Which of the following is the last step in protein synthesis?

A A molecule of tRNA bonds to an amino acid in the cytoplasm.

B The stop codon binds to the ribosome, and the polypeptide is released.

C DNA unravels to expose a gene segment.

D The mRNA strand bonds to tRNA.

C. Answer the following questions.

1 What are proteins made of?

2 What role does DNA play in protein synthesis?

Mutations

Mutations are mistakes or misconnections in the duplication of the chromatin material. Mutations usually occur in the nucleus of the cell during the replication process of cell division. Some mutations are harmful to an organism, some are beneficial and some have no effect. Mutations play a significant role in creating the diversity of life on Earth today. Geneticists classify mutations into two groups: **gene mutations** and **chromosomal mutations**.

Gene mutations are mistakes that affect individual genes on a chromosome. For instance, one base on the DNA strand substitutes for another base. A substitution of bases will change the codon and, therefore, the amino acid. Consequently, the protein being synthesized may be different from what the DNA originally coded for, thus possibly affecting one or more functions within the organism. Gene mutations also occur by the insertion or deletion of nucleotides from a gene.

Chromosomal mutations are mistakes that affect the whole chromosome. There are four major categories of chromosomal mutations.

- **Duplication mutations** occur when a chromosome segment attaches to a homologous chromosome that has not lost the complementary segment. One chromosome will then carry two copies of one gene or a duplicate set of genes.
- **Deletion mutations** occur when a chromosome segment breaks off and does not reattach itself. When cell division is complete, the new cell will lack the genes carried by the segment that broke off.
- **Inversion mutations** occur when a segment of chromosome breaks off and then reattaches itself to the original chromosome, but backwards.
- **Translocation mutations** occur when a chromosome segment attaches itself to a nonhomologous chromosome.

Mutations in the somatic cells affect only the tissues of the organism. Mutations occurring in the reproductive cells may be transmitted to the gametes formed in meiosis and thus pass on to future generations. These mutations sometimes cause abnormal development and a variety of genetic diseases. In other cases, the mutation benefits the organism. Many times, the mutation has little or no noticeable impact on the organism.

Although mutations do occur spontaneously, environmental factors can increase the likelihood of developing mutations. These types of environmental factors are called **mutagens**. Radiation exposure through X-rays or UV light can alter DNA in humans. Exposure to large amounts of radiation, like the kind following a nuclear bomb, can adversely affect all cells within the human body, including sex cells. This increases the statistical probability of developing mutations in organisms. Japanese survivors of the atom bomb have shown a higher than normal cancer and birth defect rate. Natural mutation-causing chemicals in food, cookware, human-made chemicals and pollutants can also cause mutations. Extremely high temperatures and some kinds of viruses can also cause mutations.

Section Review 3: Mutations

A. Define the following terms.

mutation	mutagens	chromosomal mutation	inversion mutation
gene mutation	duplication mutation	deletion mutation	translocation mutation

B. Choose the best answer.

1 What is a change in the chromosome structure caused by radiation, chemicals, pollutants or during replication?

A mutation B allele C gene D replicator

2 Which of the four types of mutations cause a change in the arrangement, rather than the number, of genes on a chromosome?

A deletion
B deletion and translocation
C translocation
D translocation and inversion

3 What type of mutation affects individual genes?

A chromosomal mutations
B gene mutations
C crossing over
D meiosis

4 When people spend a lot of time sunbathing and rarely put on sunscreen, they have an increased risk of developing skin cancer. Why is this ***true***?

A because lactic acid buildup alters their DNA in a harmful way

B because UV light from the Sun alters their DNA in a harmful way

C because the extremely high temperatures experienced while sunbathing alters their DNA in a harmful way

D because their skin becomes too dry

C. Complete the following exercises.

1 Describe how mutations are passed on to future generations.

2 Identify mutagens found in your area. Research how to limit your exposure to each type of mutagen and present your findings in a table. Some good places to start your search include air pollution reports put out by the EPA or city/county governments. Also the local water-quality report is another good starting point.

Chapter 5 Review

1 In the DNA molecule, guanine pairs with another base called

A quinine. B riboflavin. C cytosine. D thymine.

2 What is the term for sections of DNA that resemble rungs on a ladder?

A genetic codes C base pairs

B reprocessors D lipid pairs

3 Which molecule transports the code of information from the nucleus to the ribosome?

A tRNA B rRNA C mRNA D an amino acid

4 During translation, adenine on mRNA will pair with which base on tRNA?

A uracil B guanine C thymine D cytosine

5 What is the function of a stop codon?

A to instruct tRNA to stop delivering amino acids to mRNA

B to instruct the ribosome to stop delivering amino acids to mRNA

C to instruct the ribosome to stop the translation process and release the protein

D to instruct the ribosome to stop the transcription process and release the protein

6 Examine the base segment below:

AAC CTC GCG GGG

What type of biomolecule is depicted?

A DNA B RNA C amino acid D fatty acid

7 Which factor below will most likely create a genetic mutation?

A UV radiation B transfer RNA C amino acid D anticodon

8 What is the ultimate goal of protein synthesis?

A to create DNA segments C to create proteins

B to create cellular energy D to create mRNA

9 What sequence of amino acids does the following mRNA segment code for?

AUG CUA GUG AAA

A	methionine	leucine	alanine	asparagine
B	methionine	leucine	valine	lysine
C	methionine	isoleucine	lysine	tyrosine
D	methionine	leucine	methionine	histidine

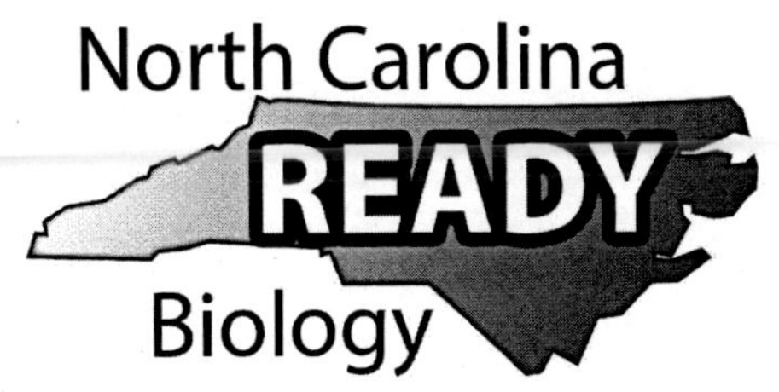

Chapter 6 Genetics

This chapter covers:
3.2.1, 3.2.2, 3.2.3

MEIOSIS

Meiosis is a type of cell division necessary for **sexual reproduction**. It is limited to the reproductive cells in the testes (the sperm cells), and the reproductive cells in the ovaries (the eggs). Meiosis produces four reproductive cells, or **gametes**. These four cells contain half the number (**haploid**) of chromosomes of the mother cell, and the chromosomes are not identical. There are two phases of cell division, **meiosis I** and **meiosis II**. Before meiosis begins, each pair of chromosomes replicates while the cell is in its resting phase (interphase).

Figure 6.1 Crossing Over of Chromosomes

During meiosis I, each set of replicated chromosomes lines up with its homologous pair. **Homologous chromosomes** are matched pairs of chromosomes. Homologous chromosomes are similar in size and shape and carry the same kinds of genes. However, they are not identical, because each set comes from a different parent. The homologous pairs of chromosomes can break and exchange segments during the **crossing over** process, a source of genetic variation. The homologous pairs of chromosomes separate. The cell then splits into two daughter cells, each containing one pair of the homologous chromosomes. **Interkinesis** is the resting period before meiosis II begins.

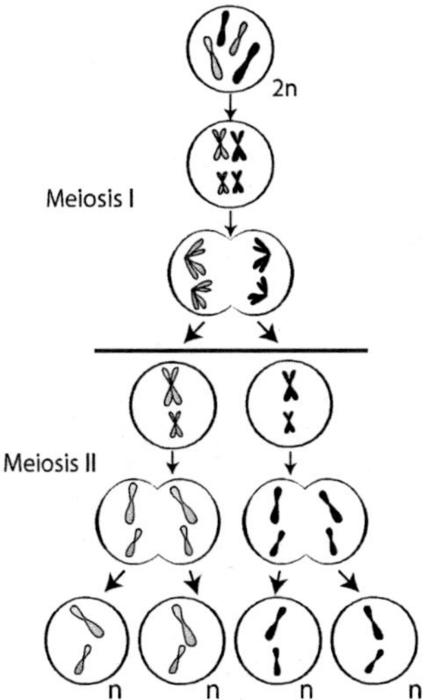

Figure 6.2 Meiosis

During meiosis II, the two daughter cells divide again without replication of the chromosomes. The result is four gametes, each having half the number of chromosomes of the mother cell.

In human males, all 4 gametes each produce a long whip-like tail, now called sperm. In human females, 1 gamete forms an egg cell with a large supply of stored nutrients. The other 3 gametes, called polar bodies, disintegrate.

In humans, the body cells have 23 different pairs or a diploid (2n) number of 46 chromosomes total. Each egg and each sperm have 23 single or haploid (n) number of chromosomes.

The process of meiosis is vital to maintaining the correct number of chromosomes given to each new offspring. Errors in meiosis result in too few or too many chromosomes in the offspring cell. Several human diseases are attributed to incorrect chromosomal number, including Down's syndrome, Turner syndrome and Klinefelter's syndrome. A mistake in meiosis can have far-reaching consequences for offspring organisms.

Asexual vs. Sexual Reproduction

Asexual reproduction by mitosis is a careful copying mechanism. Some unicellular organisms, like amoeba, produce asexually. Many plants also produce asexually. There are several mechanisms by which this occurs. However, the offspring produced are always genetically identical to the parent.

In contrast, **sexual reproduction** by meiosis brings with it the enormous potential for genetic variability. The number of possible chromosome combinations in the gametes is 2^n, where n is the haploid chromosome number and 2 is the number of chromosomes in a homologous pair. Look at Figure 6.3, which shows the possible distribution of chromosomes into homologous pairs at meiosis in organisms with small numbers of chromosomes, in this case 2 and 3.

When n = 2, four distinct distributions are possible. When n = 3, eight distinct distributions are possible. If humans have a haploid number of n = 23, then 2^{23}, or 8,388,608 distinct distributions are possible. Remember, this is only the genetic variation that occurs *before* fertilization. After fertilization, the genetic variation only increases.

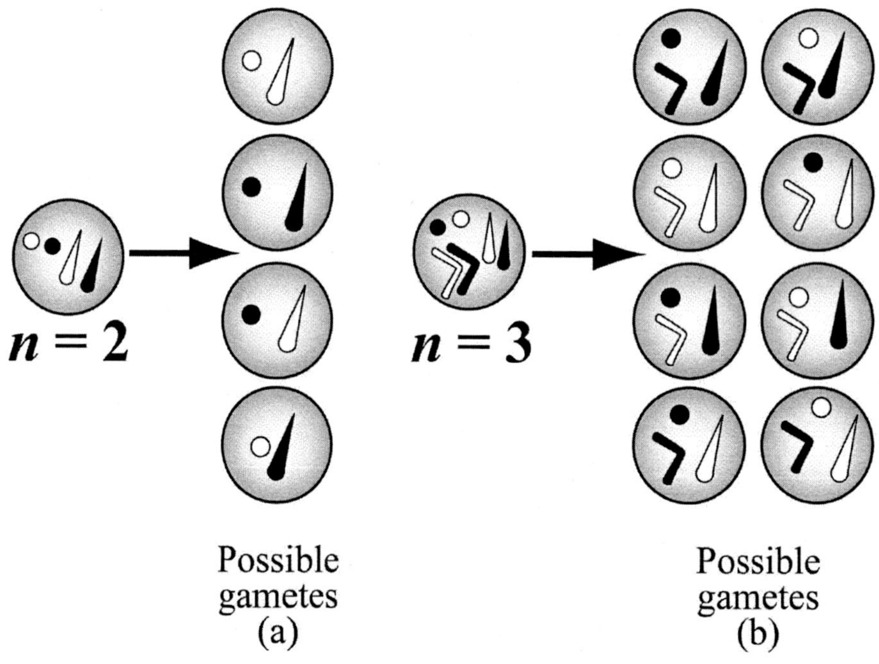

Figure 6.3 Genetic Variability

Section Review 1: Meiosis

A. Define the following terms.

reproductive cells	chromatin	telophase	sexual reproduction
haploid	asexual reproduction	cytokinesis	gamete
somatic cells	prophase	cell division	crossing over
diploid	replication	meiosis	interkinesis
homologous chromosomes			polar bodies

B. Choose the best answer.

1 What is a common term for a sperm or an egg cell?

A gamete B interkinesis C somatic D spindle

2 Which type of nuclear division produces gametes?

A meiosis B cytokinesis C interphase D mitosis

3 How does sexual reproduction ***most*** benefit species?

A It increases mutation.

B It increases genetic variation.

C It decreases number of chromosomes.

D It decreases diploid chromosomes.

4 How does meiosis ***most*** differ from mitosis?

A age of cells initially used

B length of chromosomes

C number of organelles in daughter cells

D amount of possible genetic variations

5 If a plant cell has 10 chromosomes, how many possible genetic distributions can occur during meiosis?

A 100 B 200 C 1024 D 2048

C. Complete the following exercise.

1 The normal number of chromosomes in a yellow pine tree is 24. With pictures taken from a high-powered microscope, you determine that the pollen from the yellow pine only has 12 chromosomes. How can this be explained?

2 List and describe the two main ways meiosis differs from mitosis.

GENETIC EXPRESSION

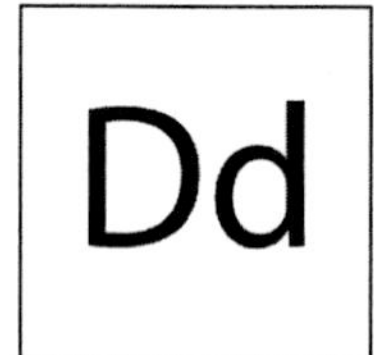

Figure 6.4 Genotype for Dimples

Genes, which are specific portions of DNA, determine hereditary characteristics. Genes carry traits that can pass from one generation to the next. **Alleles** are different molecular forms of a gene. Each parent passes on one allele for each trait to the offspring. Each offspring has two alleles for each trait. The expression of physical characteristics depends on the genes that both parents contribute for that particular characteristic. **Genotype** is the term for the combination of alleles inherited from the parents.

Figure 6.5 Phenotype for Having Dimples

Genes can be either dominant or recessive. The **dominant gene** is the trait that will most likely express itself. If both alleles are dominant, or one is dominant and one is recessive, the trait expressed will be the dominant one. In order for expression of the **recessive gene** to occur, both alleles must be the recessive ones. For example, a mother might pass on a gene for having dimples, and the father might pass on a gene for not having dimples. Having dimples is dominant over not having dimples, so the offspring will have dimples even though it inherits one allele of each trait. For the offspring not to have dimples, both the mother and father must pass along the allele for not having dimples. The **phenotype** is the physical expression of the traits. The phenotype does not necessarily reveal the combination of alleles.

When studying the expression of the traits, geneticists use letters as symbols for the different traits. Capital letters are used for dominant alleles and lowercase letters for recessive alleles. For dimples, the symbol could be D. For no dimples, the symbol could be d. The genotype of the offspring having one gene for dimples and one gene for no dimples is Dd. Figure 6.4 shows an example of a genotype. The phenotype for this example is having dimples. Figure 6.5 shows the phenotype of having dimples.

If an individual inherits two of the same alleles, either dominant or recessive, for a particular characteristic, the individual is **homozygous**. If the offspring inherits one dominant allele and one recessive allele, such as in the example in the above paragraph, the individual is **heterozygous**. Table 6.1 helps to clarify these terms.

Table 6.1 Alleles for Dimples

homozygous	DD, dd
heterozygous	Dd

Activity

Give all possible genotypes and phenotypes for each problem below.

	Genotypes	Phenotypes
1 Black fur is dominant over white fur.		
2 Long legs are dominant over short legs.		
3 Jointed appendages are dominant over straight ones.		
4 Oval bodies are recessive to round bodies.		
5 Round eyes are dominant over almond eyes.		
6 Webbed feet are dominant over non-webbed feet.		
7 Fur-tipped ears are recessive to smooth ears.		
8 Straight fins are dominant over curved fins.		
9 Twisted feathers are recessive to straight feathers.		
10 Ten digits are recessive to twelve digits.		

Genetics

Geneticists use the **Punnett square** to express the possible combinations for a certain trait an offspring may inherit from the parents. It is a model that helps to show the statistical probability of the traits found in offspring. The Punnett square shows possible genotypes and phenotypes of one offspring. Figure 6.6 below shows an example of a **monohybrid cross**, which involves one trait, done on a Punnett square.

The Punnett Square

The Punnett square is a tool geneticists use to determine the possible genotype of one offspring. The possible alleles donated by one parent are written across the top and the possible alleles donated by the other parent are written along the left side. In the example, the cross between two heterozygous parents is examined.

D = allele for dimples

d = allele for no dimples

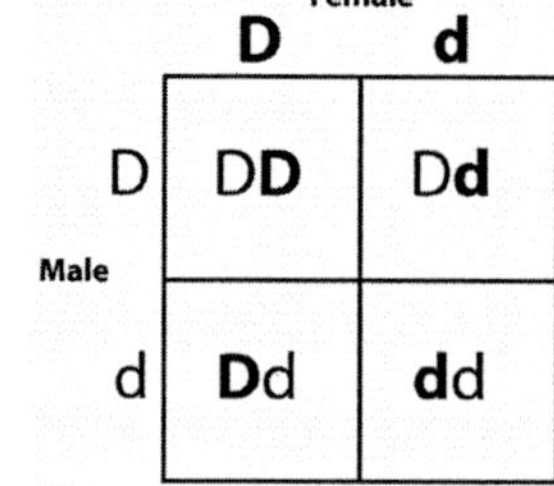

Each time this male and female produce an offspring, there is a 3/4 (or 75%) chance the offspring will have dimples and a 1/4 (or 25%) chance the offspring will have no dimples.

Figure 6.6 Punnett Square for Dimples/No Dimples

The phenotype depends not only on which genes are present, but also on the environment. Environmental differences have an effect on the expression of traits in an organism. For example, a plant seed may have the genetic ability to have green tissues, to flower and to bear fruit, but it must be in the correct environmental conditions. If the required amount of light, water and nutrients are not present, those genes may not be expressed. In another example, researchers discovered that if they kept fish embryos in a strong solution of magnesium chloride, then the genes that code for number of eyes are affected. High levels of magnesium chloride produced fish with only one eye.

Temperature also affects the expression of genes. Flowering primrose plants will bloom red at room temperature and white at higher temperatures. Himalayan rabbits and Siamese cats have dark extremities like ears, nose and feet, at low temperatures. Warmer areas of the animals' bodies are lighter colored.

There are many examples of environmental factors affecting the expression of human genes. One simple example includes sunlight exposure. Exposure to sunlight helps the human body utilize vitamin D and folic acid. However, excessive sunlight exposure leads to skin cancer. Tobacco use is another example of an environmental condition affecting the expression of genes. It is a well-established fact that using tobacco will cause cancer. Chemicals in the tobacco interfere with genes that control cell cycling. Over time, these chemicals eventually allow uncontrolled cell division to happen, causing lung or mouth cancer. Adult-onset or Type II diabetes is a third example. Consuming large amounts of refined sugars and getting little exercise causes spikes in blood sugar. Over time, the body's natural sugar carrier, insulin, no longer functions properly.

Mendel's Contribution to Genetics

Around 1850, **Gregor Mendel** (1822 – 1884) began his work at an Austrian monastery. Many biologists call Mendel "the father of genetics" for his studies on plant inheritance. Mendel and his assistants grew, bred, counted and observed over 28,000 pea plants.

Pea plants are very useful when conducting genetic studies because the pea plant has a very simple genetic makeup. It has only seven chromosomes, its traits can be easily observed, and it can **cross-pollinate** (have two different parents) or **self-pollinate** (have only one parent). Table 6.2 lists some of the pea plant traits, along with their attributes. To begin his experiments, Mendel used plants that were true breeders for one trait. **True breeders** have a known genetic history and will self-pollinate to produce offspring identical to the parent.

Table 6.2 Possible Traits of Pea Plants

Trait	Attributes	Trait	Attributes
Seed Shape	Round* Wrinkled	Pod Color	Green* Yellow
Seed Color	Yellow* Green	Flower Position	Axial* Terminal
Seed Coat Color	Gray* White	Plant Height	Tall* Short
Pod Shape	Smooth* Constricted		

*Dominant

Principle of Dominance

Through his experiments, Mendel discovered a basic principle of genetics, the principle of dominance. Mendel's **principle of dominance** states that some forms of a gene or trait are dominant over other traits, which are called recessive. A dominant trait will mask or hide the presence of a recessive trait. When Mendel crossed a true breeding tall pea plant with a true breeding short pea plant, he saw that all the offspring plants were tall. The tallness trait *masks* the recessive shortness trait. Sometimes recessive traits can be hidden in several generations, reappearing in future offspring. The crossing of the true breeders is the **parental generation**, or the **P** generation. The offspring produced are the first filial generation or $\mathbf{F_1}$ generation. The offspring of the F_1 generation are called the second filial or $\mathbf{F_2}$ generation.

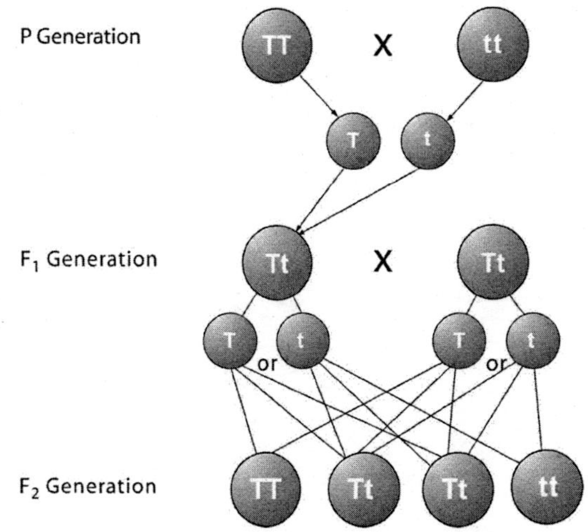

Figure 6.7 Possible Genotypes of Offspring

Principle of Independent Assortment

When Mendel began to study **dihybrid crosses**, which involve two traits, he noticed another interesting irregularity. Mendel crossed plants that were homozygous for two traits, seed color and seed texture. Round seed texture and green color are both dominant traits. Mendel assigned the dominant homozygous P generation the genotype of (RRGG). Wrinkled seed texture and yellow color are both recessive traits. The recessive homozygous P generation seeds were assigned the genotype (rrgg). When (RRGG) was crossed with (rrgg) the resulting F_1 generation was entirely heterozygous (RrGg). The F_1 generation was then allowed to self-pollinate, resulting in an F_1 dihybrid cross of (RrGg) with (RrGg). The result was an F_2 generation with a distinct distribution of traits, as depicted in Figure 6.8. Counting up the genotypes of the F_2 generation should give you the result that 9/16 of them will have the round, green phenotype, 3/16 will have the round, yellow phenotype, 3/16 will have the wrinkled, green phenotype and 1/16 will have the wrinkled, yellow phenotype.

Dihybrid Cross

	RG	Rg	rG	rg
RG	RRGG	RrGg	RrGG	RrGg
Rg	RRGg	RRgg	RrGg	Rrgg
rG	RrGG	RrGg	rrGG	rrGg
rg	RrGg	Rrgg	rrGg	rrgg

Figure 6.8 Dihybrid Cross of RrGg × RrGg

The consistent observation of this trend led to the development of **the principle of independent assortment**. This principle states that each pair of alleles segregates independently during the formation of the egg or sperm. For example, the allele for green seed color may be accompanied by the allele for round texture in some gametes and by wrinkled texture in others. The alleles for seed color segregate independently of those for seed texture.

This is where the true beauty of meiosis becomes apparent. Because of the way the chromosomes segregate into the various daughter cells, traits also randomly and independently move to offspring. Traits that are located on different chromosomes pass to different offspring. This explains how siblings can look similar but not identical to each other.

Activity

Fill in the Punnett squares below with the given genotypes to determine the possible phenotypes of the offspring.

1. GG × Gg

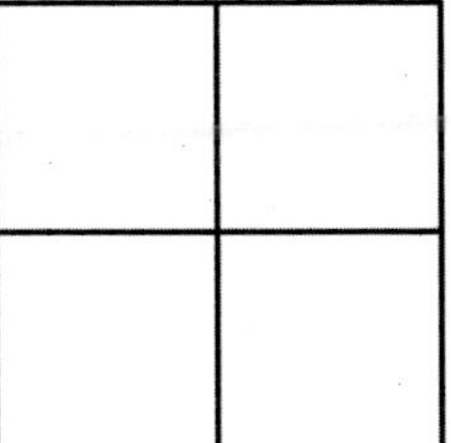

2. Jj × jj

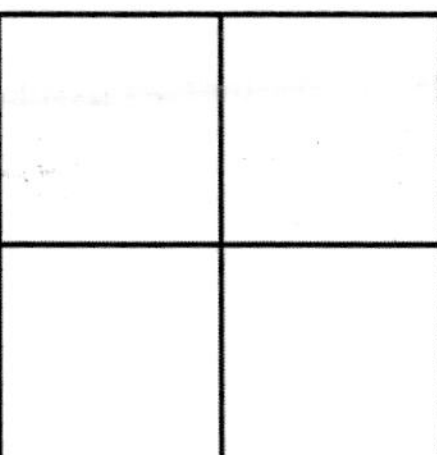

3. pp × PP

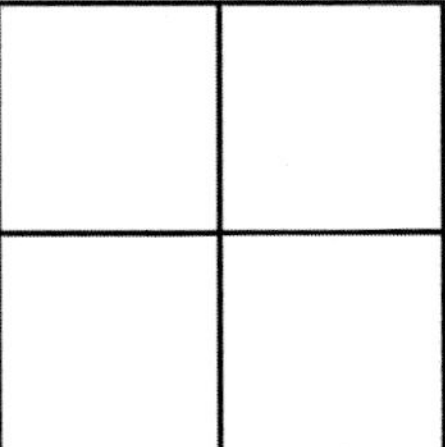

4. Kk × Kk

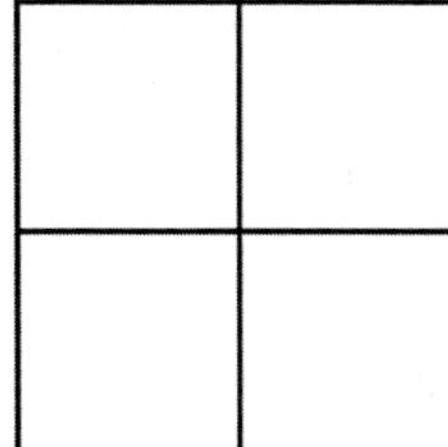

5. qq × QQ

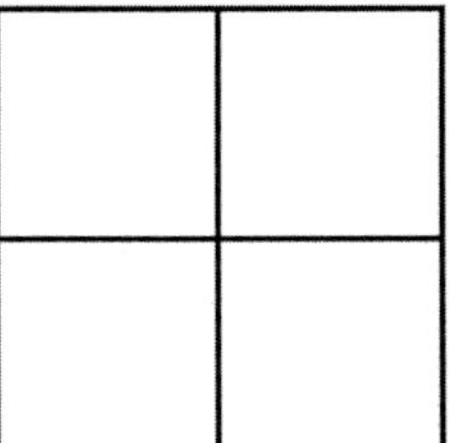

6. Hh × HH

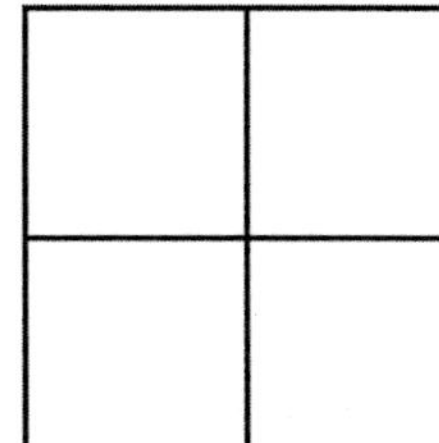

7. Cc × cc

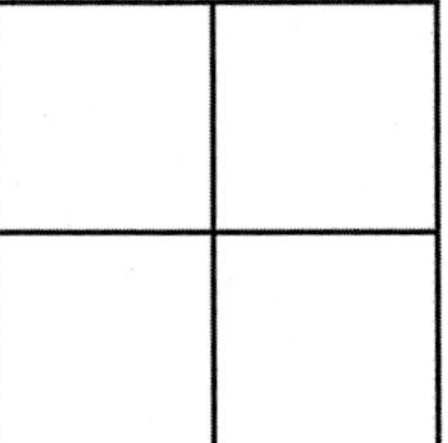

8. Ee × Ee

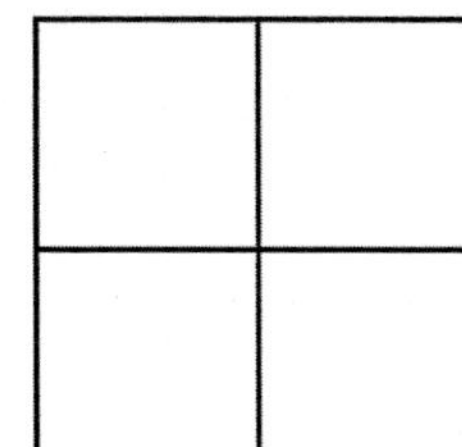

9. OoDd × OoDd

10. HhGg × HHGg

Section Review 2: Genetics

A. Define the following terms.

gene	phenotype	Gregor Mendel
allele	homozygous	true breeder
genotype	heterozygous	principle of dominance
dominant gene	Punnett square	principle of segregation
recessive gene	monohybrid cross	dihybrid cross
self-pollinate	cross-pollinate	principle of independent assortment

B. Choose the best answer.

1 What is the combination of inherited alleles called?
A heterozygote
B phenotype
C genotype
D Punnett square

2 What is the physical expression of traits called?
A phenotype
B genotype
C mutation
D allele

3 If an individual inherits one dominant allele and one recessive allele, what is the genotype?
A homozygous
B recessive
C heterozygous
D phenotype

4 If an individual inherits two of the same allele, either both dominant or both recessive for a particular characteristic, what is the individual's genotype?
A heterozygous
B phenotypic
C homozygous
D mutated

5 Use a Punnett square to predict the cross of a homozygous green parent with a homozygous yellow parent if yellow is dominant over green. What will the phenotype of the offspring be?
A all yellow
B all green
C neither yellow nor green
D some yellow and some green

C. Answer the following questions.

1 What is the relationship between phenotype and genotype?

2 Compare homozygous alleles to heterozygous alleles.

Modes of Inheritance

Up to this point we have been discussing organism inheritance in terms of **classical genetics**. Classical genetics utilizes ideas on heredity that predate the discovery of DNA as the molecule of inheritance. In classical genetics, most ideas on organism inheritance were developed without any idea of the molecular basis of DNA. Classical genetics relies mostly on Mendel's laws. In Mendel's time, humans had no idea how traits were passed from one generation to the next. He had no idea about chromosomes, genes, alleles or the process of meiosis.

Figure 6.9 Watson

In contrast, the discipline of **molecular genetics** recognizes DNA as the hereditary material of an organism. Molecular genetics understands that chromosomes are made up of long strands of DNA. DNA is divided into segments called genes. The structure of DNA was first defined by **James D. Watson** and **Francis Crick** in their paper published in 1953. Watson and Crick used experimental results obtained by Rosalind Franklin to help build their famous model of DNA. It is now a widely known fact that DNA and its specific order of nucleotide segments determines the proteins an organism can produce. It is these proteins that alter the look and function of organisms.

Figure 6.10 Crick

As it turns out, not all traits are strictly dominant or recessive as postulated by Mendel. The extremely complex structure and interaction of genes and alleles makes heredity a hot topic for research. In addition, the Human Genome Project and modern sequencing techniques give molecular geneticists insight into the genetics of organisms. As we will soon see, there are special circumstances where traits are inherited in a more complex way. Some traits are coded on special sex chromosomes that are unevenly divided between males (XY) and females (XX). Other traits are controlled by the interaction of several genes and, to some extent the environment.

Sex-Linked Traits

Sex chromosomes are the chromosomes responsible for determining the sex of an organism. These chromosomes carry the genes for sex determination as well as other traits. Thomas Hunt Morgan was the first to discover sex-linkage in 1907. Sex chromosomes are the 23rd pair of chromosomes and are sometimes called X or Y chromosomes. Males have the genotype XY and females have the genotype XX. In females, one X comes from the mother and one X comes from the father. In males, the X chromosome comes from the mother and the Y chromosome comes from the father.

If a recessive trait, like color blindness, is located on the X chromosome, it is not very likely that females will have the phenotype for this condition. Remember females have two X chromosomes. If one X chromosomes contains the recessive trait, the alternative X chromosome might contain the dominate trait. It is more likely that males will have the condition since they only have one X chromosome, whereas females have two. Males do not have another X chromosome or a duplicate copy of the gene. A female that has a recessive gene on one X chromosome is a **genetic carrier** for that trait. A carrier usually is not affected with the condition themselves, however they are able to pass the disease along to their offspring. Due to the recessive nature of the trait, carriers show no signs or symptoms of the disease. It is important to note that genetic carriers can have recessive disease mutations on the sex chromosomes OR they can have recessive disease mutations on an autosomal (nonsex) chromosome. However, autosomal recessive diseases are rare.

Punnett Square for Color Blindness

	X^B	X^b
X^B	X^BX^B	X^BX^b
Y	X^BY	X^bY

B = Normal
b = Color Blind

Figure 6.11 Punnett Square for Color Blindness

Examine the Punnett square in Figure 6.11, which shows the cross of a female who is heterozygous for color blindness with a normal male. This Punnett square shows how a mother contributes to the color blindness of her sons. Notice from Figure 6.11 that this mother has a fifty percent chance of producing a colorblind son; whereas, there is a zero percent chance of producing a colorblind daughter.

Hemophilia is another sex-linked disorder. This is a blood disorder that causes excessive bleeding in affected individuals. This disease was prominent in European royalty during the 1900's.

INCOMPLETE DOMINANCE

Incomplete dominance, sometimes called intermediate dominance, is the situation where one trait is not completely dominant over the other. Think of it as blending of the two traits. All of the offspring in the F_1 generation will show a phenotype that is a blending of both the parents. If the F_1 generation is self-pollinated, the ratio of the offspring will appear in a predictable pattern. One offspring will look like one parent, two offspring will look like both parents and one offspring will look like the other parent.

A cross between a red and a white four o'clock flower demonstrates this point. One flower in the parental generation is red with genotype R^1R^1. The other flower is white with genotype R^2R^2. The offspring of this cross appear pink and have a genotype of R^1R^2. See Figure 6.12 for the genotypes and the phenotypes of the P, F_1 and F_2 generations.

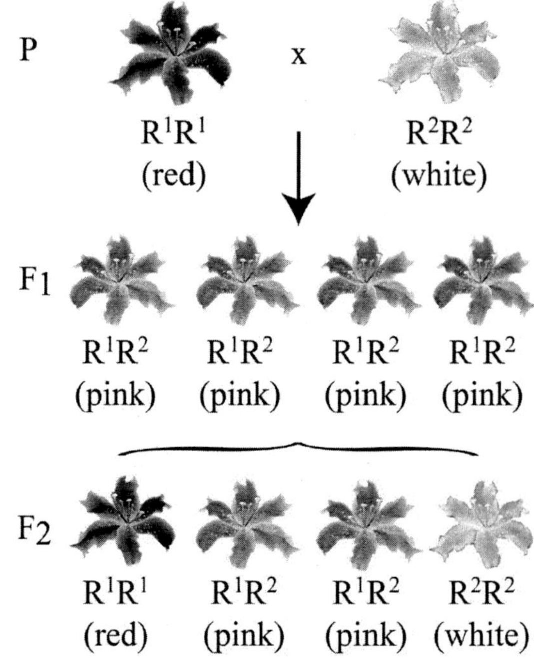

Figure 6.12 Genotypes and Phenotypes of P, F_1, and F_2 Generations of Four O'clock Flower

Activity

Research the prevalence of hemophilia in European royalty. Make a chart to show affected and unaffected individuals within the family lines.

Codominance

Figure 6.13 Checkerboard Chicken

When both traits appear in the F_1 generation and contribute to the phenotype of the offspring, the trait is **codominant**. One example occurs in chickens in which the trait for black feathers is codominant with the trait for white feathers. A checkerboard chicken has both traits, black feathers and white feathers. If you look at this bird, it has feathers with solid white and solid black patches. This makes a checkerboard chicken a good example of codominance.

Through they sound similar, there are two main differences between the situations of codominance and incomplete dominance. When one allele is incompletely dominant over another, the blended result occurs because *neither allele is fully expressed.* That is why the F_1 generation four o'clock flower is a *totally different color* (pink). In contrast, when two alleles are codominant, *both alleles are completely expressed.* The result is a combination of the two, rather than a blending. The chicken feathers are a unique combination of distinct white and black patches.

Multiple Alleles and Polygenic Traits

Certain traits like blood type, hair color and eye color are determined by two or more genes for every trait, one from each parent. Whenever there are different molecular forms of the same gene, each form is called an **allele**. Although each individual only has two alleles, there can be many different combinations of alleles in that same population. For instance, hamster hair color is controlled by one gene with alleles for black, brown, agouti (multicolored), gray, albino and others. Each allele can result in a different coloration.

Polygenic traits are the result of the interaction of multiple genes. It is commonly known, for instance, that high blood pressure has a strong hereditary linkage. The phenotype for hypertension is not, however, controlled by a single gene that lends itself to elevating or lowering blood pressure. Rather, it is the result of the interaction between one's weight (partially controlled by one or more genes), one's ability to process fats and cholesterol in particular (several metabolic genes), one's ability to process and move various salts through the bloodstream (transport genes) and one's lifestyle habits such as smoking and drinking (which may or may not be the result of the expression of several genes that express themselves as behavior types). Of course, each of the genes involved may also have multiple alleles, which vastly expands the complexity of the interaction.

Activity

Often color is inherited in an incomplete way. Make your own image chart similar to Figure 6.12, using your favorite flower and favorite color. Alternatively, make a chart showing hair color of members of your family. Include siblings, aunts, uncles, parents and grandparents.

Section Review 3: Modes of Inheritance

A. Define the following terms.

sex chromosomes	carrier	incomplete dominance
codominance	multiple alleles	polygenic traits

B. Choose the best answer.

1 A male has the genotype XY. Which parent is responsible for giving the son the Y chromosome?

A mother
B father
C both the father and the mother
D neither the father nor the mother

2 What is the difference between codominance and incomplete dominance?

A Codominant traits are blended, and incompletely dominant traits appear together.
B Codominant traits are recessive, and incompletely dominant traits appear together.
C Codominant traits appear together, and incompletely dominant traits are blended.
D Codominant traits are recessive, and incompletely dominant traits are blended.

3 A cross between a black guinea pig and a white guinea pig produces a grayish guinea pig. What information do you need to determine if the production of a grayish offspring is a result of codominance?

A the phenotype of the guinea pig's littermates
B the number of alleles per gene
C the genotype of both parents
D the type of sex-linked traits in guinea pigs

C. Complete the following exercises.

1 The phenotype for blood type is an example of a multiple allele trait. The three alleles are A, B and O — A and B are codominant, and O is recessive to BOTH A and B. Determine the phenotypes of the offspring in each of the situations shown.

(A)

Parents: AO × AA

	A	O
A	AA	AO
A	AA	AO

Offspring Phenotypes:
____ Type A, ____ Type B,
____ Type AB, ____ Type O

(B)

Parents: BO × AB

	B	O
A	AB	AO
B	BB	BO

Offspring Phenotypes:
____ Type A, ____ Type B,
____ Type AB, ____ Type O

(C)

Parents: AO × BO

	A	O
B	AB	BO
O	AO	OO

Offspring Phenotypes:
____ Type A, ____ Type B,
____ Type AB, ____ Type O

2 Can a man who has AB blood type father a child who is O? Explain.

CHAPTER 6 REVIEW

1 Use the blank Punnett square to predict the cross of a homozygous tall parent with a homozygous short parent if tall is dominant over short. The phenotypes of the offspring will be

A all tall.

B all short.

C neither short nor tall.

D some tall and some short.

2 Use the blank Punnett square to determine the probability of two heterozygous parents producing a recessive offspring.

A 0%

B 25%

C 75%

D 100%

3 In corn, the allele for long ears is dominant to the allele for short ears. If a homozygous recessive parent was crossed with a heterozygous parent, what percentage of the offspring would show the dominant phenotype? Use the blank Punnett square to help you.

A 0%

B 25%

C 50%

D 100%

4 In a dihybrid cross between completely heterozygous parents, what ratio of offspring will show both dominant phenotypes?

A 1/16

B 3/16

C 9/16

D 1/4

5 Which trait is ***most likely*** to be inherited in an incomplete way?

A hair color

B earlobe attachment

C tongue shape

D eye shape

Answer the following questions.

A variety of pea plant has flowers that may be either purple (A) or white (a); two purple pea plants are crossed, and it is known that the genotypes of the parent plants are both heterozygous. Use this information and the information given in the Punnett square to answer Numbers 6 through 8.

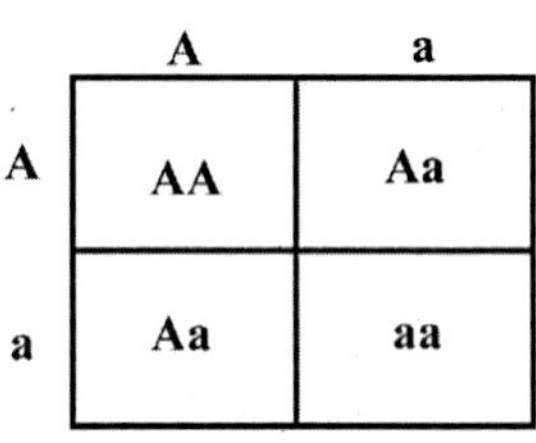

	A	a
A	AA	Aa
a	Aa	aa

6 Which trait is dominant? Which trait is recessive?

7 What percentage of the flowers will be purple? How many will be white?

8 What are the genotypes and phenotypes of the parents?

9 A man with blood type AA and a woman with blood type BB have a child. What will the blood type of their child ***most likely*** be?

A Type A

B Type B

C Type AB

D Type O

10 A child has type O blood. Which couple below ***most likely*** had this child?

A Anne has blood type A
Victor has blood type B

B Amanda has blood type O
Ray has blood type A

C Lisa has blood type O
Adam has blood type AB

D Kylee has blood type B
Chris has blood type B

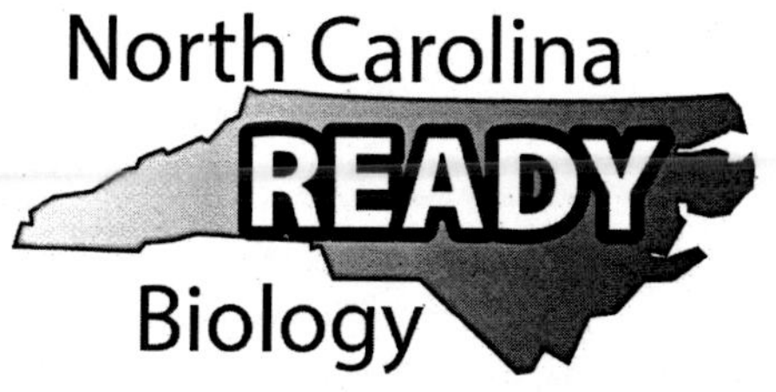

Chapter 7 Biotechnology

This chapter covers:
3.3.1, 3.3.2, 3.3.3

Figure 7.1 Biotechnology in Agriculture

Biotechnology is the commercial application of biological products and has been in existence for thousands of years. It includes the production of wine, beer, cheese and antibiotics, but today it more commonly refers to processes that manipulate DNA. DNA technologies or biotechnology manipulates DNA to benefit humans. DNA technologies have impacted many areas of modern life including medicine and agriculture. DNA technologies also have far-reaching consequences in the legal system. DNA samples are now routinely used in criminal cases. Another use of DNA technology is finding missing family members or mapping a family history. Other ways to view familial inheritance include: a graphical tool called a pedigree chart, gel electrophoresis or with a karyotype. Each method gives us different information about the inheritance pattern. Let's review these methods in greater detail.

SHOWING INHERITANCE PATTERNS

Being able to see how traits were passed down among family members allowed Mendel to develop his principles on inheritance. Familial inheritance patterns are often represented using models, pictures of chromosomes or other modern imaging techniques. Pedigrees, electrophoresis and karyotypes, are all ways scientists study inheritance patterns in organisms.

PEDIGREE CHARTS

A **pedigree** is a graphical chart used to identify the lineage of individuals. Pedigrees are useful when the genotype of individuals is unknown. Pedigrees are used in animal breeding and human families. Many times, pedigrees help show the inheritance of genetic disorders of members within families. Often males are represented with a square and females are represented with a circle. Horizontal lines found between males and females represent a mating. Males and females located at the end of a vertical line represent offspring from the above mating. Sometimes the generations found in a pedigree are numbered using roman numerals.

Members of a family affected with a genetic disease are shown as shaded or colored shapes. Sometimes individuals who are carriers for the disease are shown half shaded. Each pedigree will have a key to help you better understand the information displayed. Examine the sample pedigree in Figure 7.2.

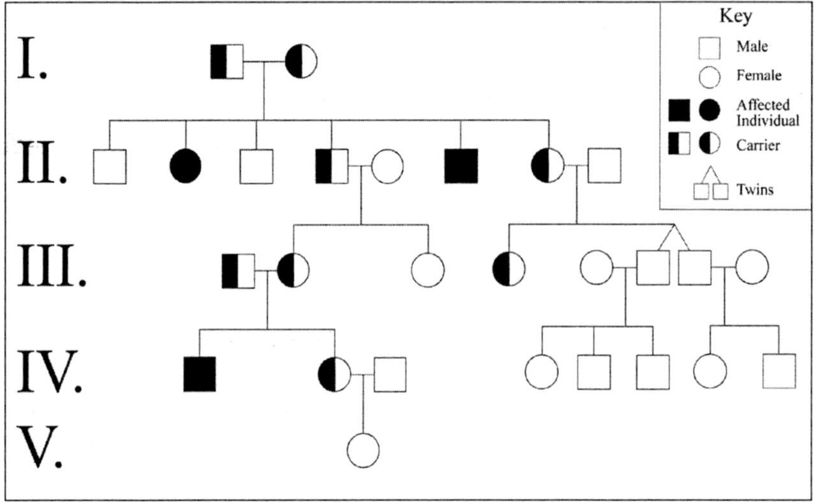

Figure 7.2 Sample of Pedigree Chart with Key

GEL ELECTROPHORESIS

One technique used to study DNA is the separation of DNA fragments using gel electrophoresis, also sometimes called DNA fingerprinting. In **gel electrophoresis**, an electric current is passed through a semisolid substance called gel. The gel substance is chosen specially for the type of molecule to be analyzed. Gel electrophoresis is used to analyze DNA and determine parentage of children or suspects in a crime. A DNA sample from, say, a crime scene can be used to compare to a sample DNA from a possible suspect. It cannot be used to tell who the person is, independent of a comparative sample. In other words, there has to be a known DNA sample to compare to the unknown DNA collected.

To begin, an unknown DNA sample is collected and cut into fragments using special enzymes. These enzymes target specific portions of the DNA molecule that is known to be very different among individuals within the population. It is an analysis of a very small portion of the DNA sample. Because of the high variation in the analyzed portion of DNA, this process is greater than 99% accurate.

The small DNA fragments are then loaded into wells at the top of the gel medium. An electric current separates and draws the negatively charged DNA fragments toward a positive terminal on the other end of the apparatus. The consistent negative charge of the DNA molecules (given by the phosphate backbone) causes the fragments to move across the gel toward the positive terminal. The speed of the fragments as they move through the gel is determined by their size and charge. Large molecules will move slowly, while small molecules will move quickly. The result is a column of bands, with each band representing a specific fragment of DNA. Each column in an electrophoresis gel represents a different DNA sample. The electrophoresis gel is often stained to make the bands visible. Sometimes a picture is taken of the gel.

Since two DNA samples will fragment and migrate in exactly the same way, identical bands represent identical segments of DNA. Figure 7.3 shows a summary of this process.

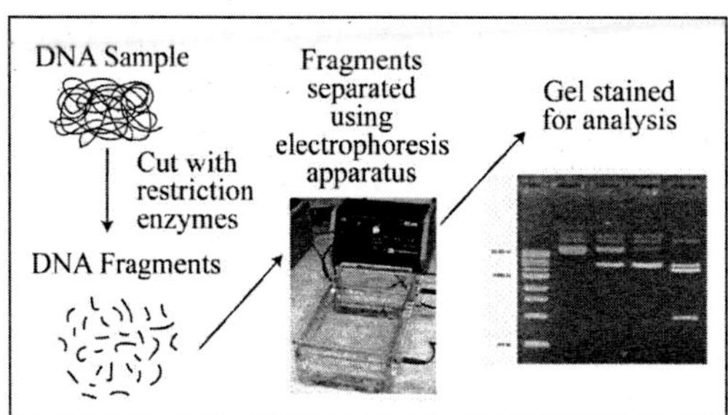

Figure 7.3 DNA Fingerprinting

Analysts can then look at the location and size of the bands to determine the relatedness of individuals who donated the DNA. The more related the individuals are, the more bands they will have in common. Identical fragments of DNA will be identical in size AND be found in an identical location within the column. Having many identical fragments indicates a close family relationship. Having few or no identical fragments indicates distant or no relatedness.

When the relationships between individuals are unknown, a DNA fingerprint can help determine their relatedness. DNA fingerprinting is often used to determine the father of a child or the suspects in a crime.

Activity

Recently, black bear populations in western North Carolina have been declining. As a result, the Department of Natural Resources (DNR) closely monitors hunters in the area. Persons caught harvesting bear meat in this area face stiff fines and possible jail time. A man named Clarence Obvious was reported to your DNR office as a possible bear poacher. As the local DNR representative, you must find him and discover if he is innocent or guilty. You get a warrant to search his freezer and discover three questionable meat samples. Clarence claims that these samples are from a deer, a boar (male hog) and a sow (female hog) he shot earlier this year. You send these samples to your DNA lab. At the lab, known samples of bear, sow, boar and deer are compared with the meat from Clarence's freezer.

Examine the results for yourself.

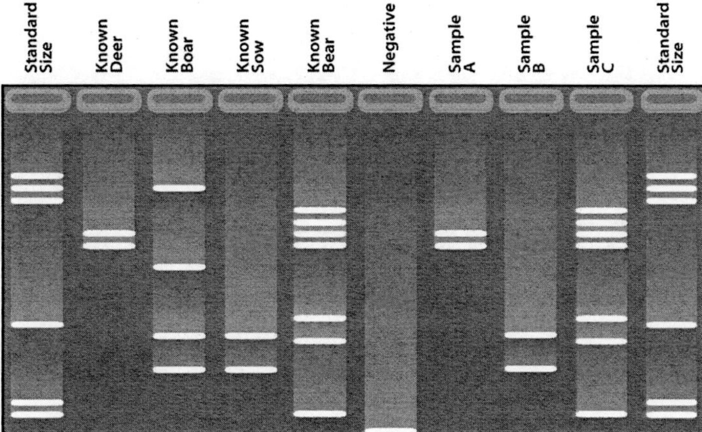

Was there bear meat in his freezer? Is Mr. Obvious innocent or guilty? How do you know?

KARYOTYPES

A **karyotype** is like a picture of the chromosomes found in the nucleus of a eukaryotic cell. It is used to examine the traits of the chromosomes. This method has been around since the early part of the 20th century. In a modern technique, called the SKY technique, specialized fluorescent molecules are attached to the chromosomes and computers are used to analyze the results. In a more traditional karyotype technique, chromosomes are extracted from a eukaryotic cell. They are then treated with a dye that gives the chromosomes a specific banding pattern. A picture of the chromosomes is taken and enlarged. Technicians identify each chromosome by its length and banding pattern. They are paired together and analyzed. Figure 7.4 shows a sample of a karyotype.

Scientists compare the overall features of the chromosomes. They examine length, shape, banding pattern and total number of chromosomes.

Karyotypes are useful in determining chromosomal disorders. For example, Trisomy 21 (Down's syndrome) is easily identified by a karyotype. The sample will contain three chromosome 21's instead of the normal two. Cri-du-Chat syndrome is also identified by a missing segment of chromosome number 5. Karyotyping is quick and fairly inexpensive compared to other techniques. It is often used to determine chromosomal errors of a fetus with a procedure called amniocentesis.

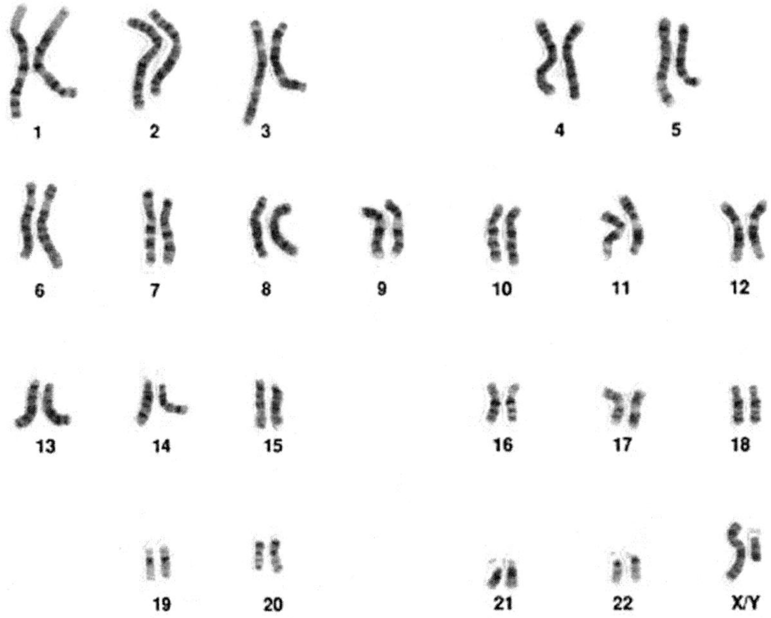

Figure 7.4 Karyotype

Genetic disorders can be identified using a pedigree gel electrophoresis or karyotype. Each technique has benefits and drawbacks, but all help us learn more about ourselves.

Section Review 1: Showing Inheritance Patterns

A. Define the following terms.

gel electrophoresis karyotype pedigree

B. Choose the best answer.

1 What determines how far a DNA fragment will travel in a gel electrophoresis apparatus?

A size

B dominance

C color

D banding pattern

2 What technique below shows inheritance without examining chromosomes or a DNA sample?

A pedigree

B electrophoresis

C DNA fingerprinting

D karyotyping

3 What does each column in a gel electrophoresis represent?

A the same individual

B a different gene

C a different sample

D parts of the same chromosome

4 What traits of chromosomes do scientists examine in a karyotype?

A banding pattern, length and overall number

B banding pattern, size and charge

C the charge on the molecule

D the movement of the DNA molecule

5 What do horizontal lines in a pedigree represent?

A offspring

B males

C females

D a mating

C. Complete the following exercise.

1 Make a concept map showing the different types of inheritance analysis.

GENETIC ENGINEERING CAN BENEFIT SOCIETY

BIOTECHNOLOGY IN MEDICINE

Figure 7.5 Biotech Analyst

The medical establishment is a strong proponent of biotechnology. The medical industry uses biotechnology to cure disease, create better drugs or to learn more about the human genome.

Tissue culturing allows bioengineers to grow human tissues and organs in the laboratory. Human ears, livers and heart cells have all been grown in the laboratory. It is hoped that these tissues can be used to help patients who are injured or need an organ transplant.

Recombinant DNA technology uses the natural process of transcription and translation to alter organisms. Recombinant DNA is often used to produce vaccines or cancer fighting medications. One way they accomplish this goal is through bacteria and viruses. These microbes can be used to target specific cells or produce specialized medicines.

Gene therapy is used by the medical community to help cure diseases. The idea is that if a defective protein is replaced with a good one, then the disease caused by the defective protein can be eliminated. Gene therapy has the greatest potential for success in treating diseases with only one defective gene. Gene therapy is used to treat SCID, severe combined immunodeficiency. However, two of the children that received gene therapy to cure their SCID later developed leukemia. This brings to light an important point: the challenges involved in gene therapy. It's not as simple as cutting out the defective gene and just pasting in the functioning one. In order for gene therapy to be successful, several challenges must be conquered. First, the functioning gene must get to the correct tissues. Then the gene must "paste" itself into the correct place within the DNA strand. Finally, the functioning gene must be turned on to begin producing the correct proteins.

The first challenge of gene therapy is a tough one. Remember, tissues are made up of thousands or millions of cells. Each one of these millions of cells must receive a copy of the gene while avoiding the body's own immune system. Visit the website http://learn.genetics.utah.edu/content/tech/genetherapy/gttools/ to learn more about how scientists introduce genes into a genome.

The newly inserted gene must then be spliced into the correct location on the DNA strand within *each cell* found in the target tissue. This was the problem with the two SCID patients who later developed leukemia. Their copies of the functioning genes "pasted" themselves in an incorrect place on the DNA strand. This disrupted other genes (that controlled cell cycle time) on the same chromosome. Gene therapy to treat SCID is successful for the majority of the children who undergo the procedure.

In perhaps the greatest challenge, gene activation, genes must be switched on while inside the target cell to begin producing the correct protein product. This topic is currently the subject of much scientific research.

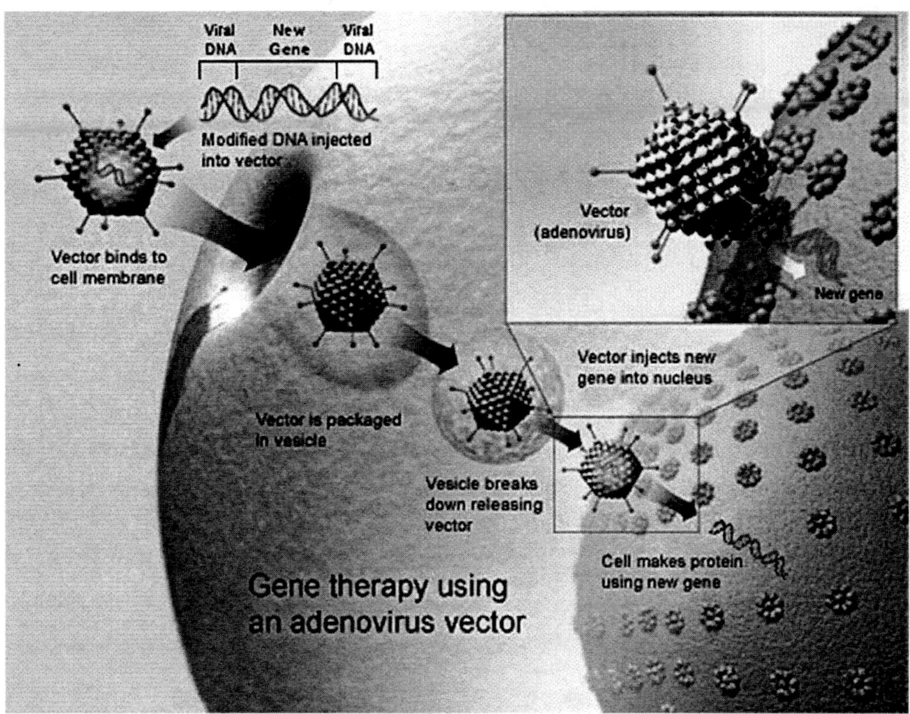

Figure 7.6 Gene Therapy Using a Virus as a Vector

Figure 7.7 Human Insulin Crystals

Biotechnology is often used in **drug development**. One such example is bioengineered bacteria that have been used to treat patients suffering from diabetes. The gene that produces human insulin has been inserted into bacterial cells. The bacteria, which reproduce quickly, can make vast quantities of insulin far cheaper than traditional animal sources of the hormone. Figure 7.7 shows human insulin crystals produced in this way. This type of medical therapy has been used to help people suffering from blood disorders, hormone imbalances, arthritis and growth deficiencies. It has also been used to produce cheap antibiotics. Using bacteria in this way is called bacterial transformation. **Bacterial transformation** uses bacteria to create beneficial products. The steps in bacterial transformation are summarized below.

1 Insert gene into the bacterial plasmid.

2 Get bacteria to take up the plasmid.

3 Grow the transformed bacteria in culture.

4 Collect the product.

BIOTECHNOLOGY OF AGRICULTURAL CROPS

Many scientists and researchers believe that recombinant technology holds great potential for improvements in agricultural products. There have already been many successes with the technology. These modified crops and animals allow farms to produce higher quality and more bountiful products, which in turn give the farmers a greater earning potential. For centuries, traditional methods of domestication have been widely used to improve the genetic characteristics of various agricultural products.

Traditional biotechnology utilizes a process called selective breeding. **Selective breeding** is the process where humans breed select individuals to pass along only desired traits to future generations. This is how humans control the physical characteristics and temperament of many domesticated plants and animals. Successive breeding of individuals showing the desired traits is continued for several generations until all offspring consistently show the desired trait.

In a simple example, a goat breeder wants to produce a goat with small horns. He need only breed the goats with the smallest horns in each generation until finally his goats only have kids with short horns. At this point, he has successfully produced a variety of goats. All future generations of this type of goat carry this deliberate mutation in horn length.

In another example, researchers in Israel have created a hybrid featherless or "naked" chicken with traditional breeding techniques. A natural mutation gave a breed of chicken a featherless neck. Selective breeding then eventually produced a completely bald chicken.

Not pretty to look at, one might wonder why chicken farmers would want to produce such an animal. The answer is simple: feathers. Much of the food and energy a chicken consumes goes into growing feathers. In hot humid climates, feathers adversely affect the growth rate and cause overheating or death. Feathers must be removed before consumption and require lots of water at the processing plant. Several advantages like reducing energy consumption used by farmers, reducing the amount of feed required, reducing the cost of chicken and reducing the amount of time to process it far outweigh its ugly exterior. The naked chicken is both environmentally and economically friendly.

Figure 7.8 Featherless Chicken

Most of the domesticated plants and animals we know today were created because they are advantageous to humans. Plants or animals became larger, easily trained, durable, performed a job, were more nutritious, tasted better and grew quickly. However, traditional selective breeding practices can take many years before they are successful.

This brings me to my next point: All new technology bears with it some risks and some benefits. Agricultural engineers must weigh the risks and benefits of each new technology. If the benefits outweigh the risks, then the technology is likely to be accepted and put into practice.

Recombinant technology takes traditional methods to an improved level by allowing scientists to transfer specific genetic material in a very precise and controlled manner and in a shorter period of time. For example, in plant crops, the characteristics of pest resistance and quick growth in harsh environments are highly desirable. In addition, improving crop quality is a long-standing goal of agricultural growers. Larger, more nutritious foods are sought using genetic engineering.

Recombinant technology has created improved strains of corn, rice, soybeans and cotton. These crops have desirable genes that enable them to resist certain insects or tolerate herbicides. This reduces the amount of chemicals, which lowers costs for the farmers, as well as helping to reduce environmental damage and runoff pollution.

Some improved products show promise for making a global impact on the problem of malnutrition. Improved strains of rice and corn contain more vitamins and can grow under stressful environmental conditions, like drought. These are valuable benefits that must be considered when discussing recombinant crops.

There are many questions about the possible long-term effects of these genetic technologies. One concern is that genetically modified foods may be detrimental to human health. Many question how the human body reacts to genetically modified foods. Some suggest that harmful biochemical changes occur at the cellular level when genetically modified foods are consumed. Genetically engineered foods may also cause unexpected **allergic reactions** in people, since proteins not naturally found in the product have been inserted. Without labeling, a person with allergies may find it difficult to avoid a known food allergen if part of the food causing the allergy is genetically added into another food product. In many European and Asian countries, modified foods must be labeled as such, but in the United States the FDA has not yet required consumer information labeling.

Genetic pollution can occur through the cross-pollination of genetically modified and nongenetically modified plants by wind, birds and insects. One method developed to combat this problem is a genetically modified seed called terminator seeds. **Terminator seeds** grow into crop plants that are incapable of producing fertile seeds. These plants cannot flower or grow fruit after the first season. This prevents their modified genome from escaping into the wild. Overall, the risk and benefits of this type of technology must be discussed.

Biotechnology in Microbiology

Microbiology is a field of science that studies microbes. Microbiologists attempt to find out how microbes interact with the plants and animals in our world. There are 2 reasons why the study of microbes is important: microbes can work for humans or against them. Bacteria in our intestines help humans absorb some vitamins and nutrients. Soil bacteria even help plants trap nitrogen, an important plant nutrient. Remember how bacteria and viruses are used in medicine? This is just one example of how microbes are helpful to humans.

Figure 7.9 Microbe Culture

More importantly, on a much larger scale, microbes are nature's garbage men. These tiny organisms spend their time processing wastes and recycling matter within ecosystems. Sometimes bacteria can easily break down waste products that are toxic to other organisms. Bacteria, protists and fungi live in soils, estuaries, freshwater and marine ecosystems. They constantly consume waste products cycling important nutrients through ecosystems. Remember how quickly bacterial cells can reproduce? Helping them reproduce faster or making bacteria that will eat harmful pollutants can help humans. They can also help humans during the clean up after an industrial accident or increase the rate of nutrient cycling through ecosystems.

Microbes are also used in the production of many food products. Yogurt, cheese, alcohol, bread, aged meats, soy and sauerkraut are all products made with microbes. Controlling the production of harmful bacteria in foods is important in food safety. Controlling microbes in water sources is just as critical. Ensuring the safety of water sources is an important field in microbiology.

Section Review 2: Biotechnology

A. Define the following terms.

biotechnology	gene therapy	cloning
recombinant DNA	stem cell	selective breeding
genetic pollution	genome	Human Genome Project

B. Choose the best answer.

1 What is the commercial application of biological products commonly called?

A illegal
B biotechnology
C unethical
D agricultural

2. How does biotechnology use viruses or bacteria?

A as a carrier to insert a desired DNA segment into an organism
B to copy a specific antigen
C to respond to and destroy cancer cells in humans
D to produce proteins for human consumption

3. Strawberries have been created to resist the harmful effects of frost. This is an application of what?

A genetic engineering
B gene therapy
C DNA fingerprinting
D cloning

4. A person with a defect in one gene that codes for a specific protein could be a candidate for which of the following?

A cloning
B DNA fingerprinting
C gene therapy
D protein injections

5. How long have humans used biotechnology to create organisms with desirable characteristics?

A only in the last 50 years
B for hundreds of years
C only in the last 10 years
D for thousands of years

C. Complete the following exercises.

1 How is genetically modified food beneficial to farmers? How can it be harmful?

2. Give an example of an advance in biotechnology that you have heard about in the news or read about in this chapter. Explain the benefits of the application of biotechnology as well as possible negative effects.

ETHICAL ISSUES IN BIOTECHNOLOGY

Many products of biotechnology have been genetically altered. The subject of genetic alteration is a topic of hot debate. Part of the debate is criticism of genetically altered products themselves.

A good example is transgenic crops. Many people fear that cross-pollination between transgenic crops and wild plants will have unforeseen consequences. Also, people are unsure how transgenic crops will affect humans. Some fear transgenic crops will cause people harm. Others question how the human body will react to these types of altered foods. Will the newly inserted genes cause new, unknown diseases in humans? How are these new products digested and broken down in the human body?

Let's revisit the idea of terminator seeds. Terminator seeds produce plants that create sterile seeds. Many in the global community feel it is unethical to sell this type of agricultural product to farmers in developing countries. Many farmers in developing counties rely on fertile seeds produced during the previous season to plant next year's crops. They cannot afford to purchase seeds every year. They often save seeds from one season to another. Some opponents of biotechnology fear terminator seeds will be disastrous to the farming industry.

Another part of the debate is ethical. It focuses on whether genetic modification is a good practice at all. A good example here is the genetic mapping, as in the Human Genome Project.

The **genome** is an organism's complete set of DNA, which carries the information needed for the production of proteins. Humans have 46 chromosomes that contain 30,000 genes made up of approximately 3 billion base pairs.

Launched in 1990, the **Human Genome Project** (HGP) sought to identify all human genes and determine all of the base pair sequences in all human chromosomes. The goal of the project was to chart variations in the sequence of base pairs in humans and to begin labeling the functions of genes. In addition to sequencing the human genome, the project also planned to sequence other organisms of vital interest to the biological field.

Scientists are hopeful that, by knowing the human genome, drugs can be designed based on individual genetic profiles, diagnoses of diseases will be improved and cures may be found for many genetic diseases.

The mapping of the genomes of individuals or groups of people brings up a lot of ethical questions. Who owns the information resulting from the project? Who gets to say what is a good use of the information, and what is a bad use of the information?

Figure 7.10 Human Genome Project

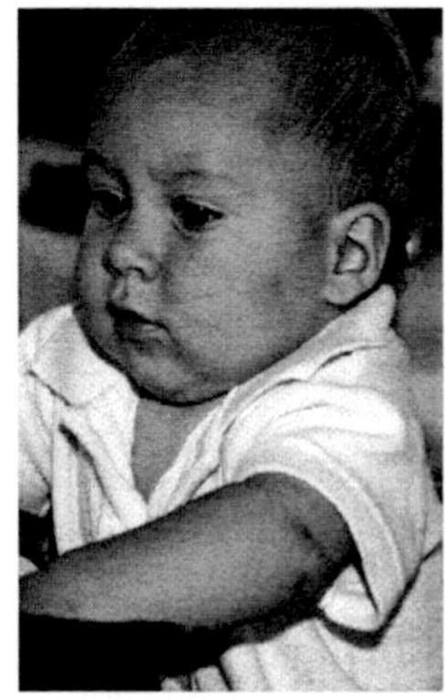

Figure 7.11
Designer Baby

And now that a genome can be mapped, it can be altered there are questions about how that technology should be used, particularly when it comes to reproduction. Is it OK to design a baby? What is to be done about genetic characteristics that are viewed as "undesirable"? What if parents want only a boy — or if they want a female child, but with blue eyes? What if they want only a genius?

In addition, just because the genome is mapped doesn't mean scientists completely understand the complex interaction between certain genes. Or, for that matter, between the environment and genes. Ultimately, what are the effects of all this genetic manipulation on complex human traits like intelligence?

The ethical questions surrounding changes in our genome are many, no matter what the application. What are the consequences? What limits should be placed on the technology? Can scientists and engineers be trusted to make solid ethical decisions? Can corporations, whose livelihood depends on increased profit make the decisions? These are hard questions, ones that require debate and research to fully understand. Let's examine two more areas where ethical choices dominate research and development: stem cell research and cloning.

Stem Cell Research

Stem cells are cells found in the human body that have yet to become a specialized type of cell. A stem cell is called a "pre cell." Stem cells have the amazing ability to become any type of cell or tissue. For example, a stem cell could develop into a nerve cell, skin cell or a liver cell. If you'll recall, we've already discussed stem cells a bit in Chapter 1 during our discussion on cell differentiation. One thing you may have missed though is that right now there are living stem cells in your body! In fact, there are *three* main sources of stem cells available. Stem cells can be harvested from adult bone marrow, umbilical cord blood after childbirth or from human embryos. The harvesting of stem cells from human embryos results in the death of that embryo. For this reason, many people oppose using embryonic stem cells in medicine. There are other avenues for harvesting stem cells, they are currently being researched as possible alternatives to the use of embryonic stem cells. These lines of stem cells show much more promising results compared to embryonic stem cells. More research is needed to determine the full range of therapeutic possibilities of stem cells. One thing is certain, the potential for using stem cells to cure disease is huge.

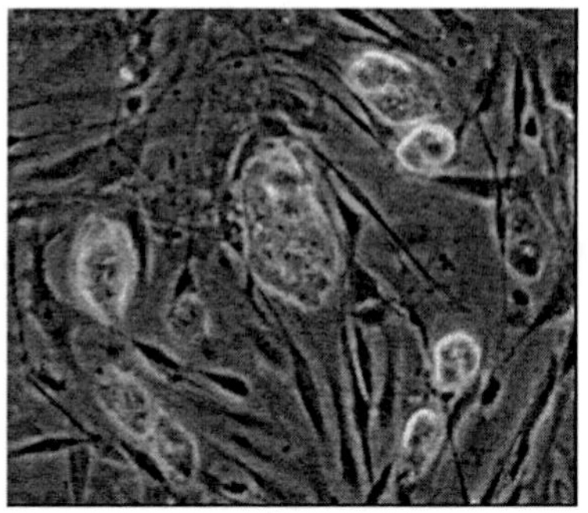

Figure 7.12 Stem Cells

CLONING

Figure 7.13 Dolly and Her Offspring Bonnie

Cloning is the creation of genetically identical organisms. In 1996, the cloning of Dolly the sheep from a somatic cell of an adult sheep created great debate about the possibility of cloning humans. The possible benefits of human cloning include enabling a childless couple to have a child, creating tissues for transplantation that would not be rejected by their host and using genetically altered cells to treat people with medical conditions such as Alzheimer's or Parkinson's, both diseases caused by the death of specific cells within the brain.

Although creating a human clone is theoretically possible, it would be very difficult. Dolly was the 277th attempt in cloning a mammal, and her death sparked a huge array of new research questions. Both scientific and moral questions must be debated, researched and solved if cloning technology is ever to become mainstream science.

Ethical questions also surround this type of technology. Some ethical questions include: Who would be allowed to create clones? Are clones considered humans, or are they the property of their original donor? Do we have the right to create hundreds of embryos for the possibility of creating a clone? Many of the implications of this type of technology remain hopeful. Clearly ethical decisions are at the forefront of this particular biotechnology.

Section Review 3: Ethics in Biotechnology

A. Define the terms.

cloning	Human Genome Project	biotechnology ethics
stem cells	terminator seeds	

B. Choose the best answer.

1 What are the three main sources of stem cells?

A karyotypes, gel electrophoresis and pedigree

B adult bone marrow, umbilical cords and embryos

C parenchymal tissues, xylem and phloem

D bone marrow, umbilical cords and phloem

2 Which source of stem cell is the ***most*** controversial?

A bone marrow

B umbilical

C phloem

D embryonic

3 What was one goal of the Human Genome Project?

A to label and identify the function of human genes

B to diagnose and cure diseases

C to create healthier and tastier food products

D to alter the human genome to create superior people

4 What is one main concern with transgenic crops?

A Who owns the rights to these crops?

B Is it moral to genetically alter animal species?

C Will these crops cross-pollinate with wild populations?

D What will happen to undesirable characteristics?

C. Complete the following exercise.

1 Have a class debate about cloning or stem cell research. Use logical arguments based on evidence.

CHAPTER 7 REVIEW

1 A police officer is at a crime scene collecting samples of blood, hair and skin. What is the officer probably going to do with the samples?

A The officer is cleaning the crime scene based on protocol.

B The officer is keeping samples to be filed with the police report.

C The officer is going to have the samples analyzed for possible DNA fingerprints.

D The officer will show them to the victim's family, the judge and the prosecutor.

2 What does each band in a gel electrophoresis represent?

A a DNA fragment

B a different DNA sample

C the charge on a DNA strand

D the electric current

3 What is used to cut DNA into fragments in a gel electrophoresis?

A enzymes

B lysosomes

C cell membrane

D ATP

4 A slight mutation on a single gene causes a rare genetic disorder. Which analyzing technique is ***most likely*** used to diagnose this disorder?

A karyotype

B pedigree

C DNA fingerprinting

D medical history

5 The medical industry is helped by technology

A through the development of better treatments and drugs.

B through the more effective disposal of wastes.

C by better helping people deal with loss of a loved one.

D through better care for the healthy members of society.

6 What is the term for genetically altered DNA?

A restricted

B fingerprinted

C recombinant

D monoclonal

Biotechnology

7 Which of the following four biomolecules will move the fastest toward the positive terminal? Circle your answer.

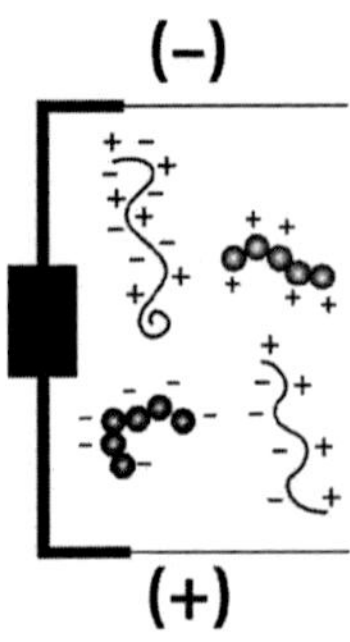

8 What is the last step in bacterial transformation?

A Grow the bacteria.

B Collect the product.

C Insert DNA into the plasmid.

D Get the bacteria to take the plasmid.

9 Which biotechnology is the oldest?

A cloning

B selective breeding

C bacterial transformation

D gene therapy

10 Examine the pedigree.

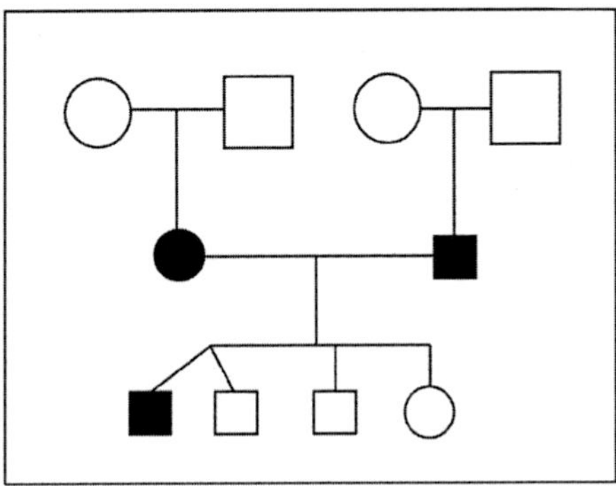

What is the mechanism by which this disorder ***most likely*** passed to offspring?

A It is recessive.

B It is dominant.

C It is sex-linked.

D It is incomplete.

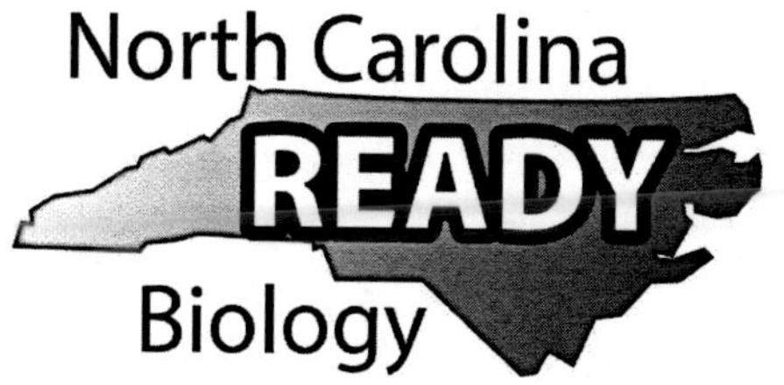

Chapter 8 Evolution

This chapter covers:
3.4.1, 3.4.2, 3.4.3

EVIDENCE OF EVOLUTION

Like every aspect of science, Darwin's ideas have been highly scrutinized. You might remember that Darwin was the first person to publish his ideas in which he proposed a mechanism by which organisms changed their physical form over time. His ideas are commonly called the theory of evolution. Many people both inside and outside the scientific community question the mechanisms proposed by Darwin. One criticism of the work states that evolution is only a theory. However, we must remember that a **scientific theory** is a well-documented explanation of natural phenomena supported by a large body of evidence. Darwin's ideas are no different. There is a vast amount of scientific evidence to suggest that organisms change over time. We will look at the specifics of Darwin's theory in the next section. Right now, we will look at the evidence used to support Darwin's ideas. Along with observed effects of natural selection, such as the long-term study of peppered moths, the natural phenomena of evolution is supported by anatomical similarities, embryonic development, genetic makeup and the fossil record.

ANATOMICAL SIMILARITIES

Anatomical similarities are evident in the study of homologous structures and vestigial organs. **Homologous structures** develop from a common ancestor and are similar in shape, but have different functions. Due to subsequent environmental changes, these body parts have very different functions. The human arm, the wing of a bird and the flipper of a whale are all homologous structures. They contain the same bones. In fact, we see a similar pattern of limb arrangement in all land-dwelling (or previously land-dwelling) organisms. The limb pattern is one bone, two bones and then many bones.

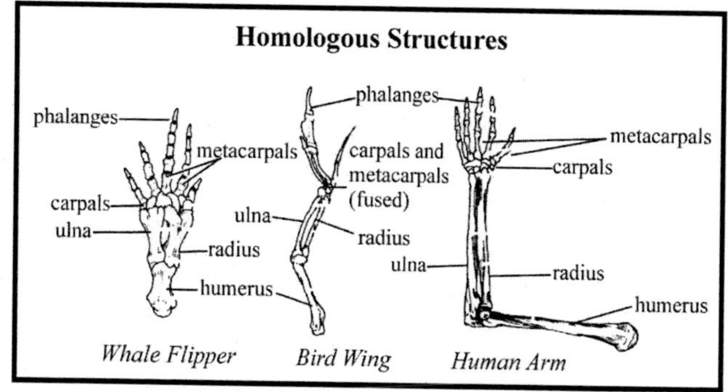

Figure 8.1 Examples of Homologous Structures

Biologists believe these structures come from a common ancestor. Their different functions correspond to their use in different environments. A whale uses a flipper for swimming, and a bird uses a wing for flying. A whale, a bird and a human all belong in the same phylum, Chordata.

Vestigial organs, structures that are no longer used or have greatly decreased in importance, are another anatomical similarity. A whale and some snakes have a pelvis and femurs, structures necessary for walking, but whales and snakes no longer have any use for these structures. These structures may have become smaller since they are unused. Their presence also suggests a common ancestor.

Activity

All mammals have 4 limbs. Each mammal limb has 5 major bone segments separated by 5 major joints. Collect images showing the skeletal structure of the forearm of a human, horse, dog and bat. Compare the skeletal structures of the bones in the forearms. Label each joint and bone with names from human anatomy. When you are finished, your collage should be similar to Figure 8.1. Some terms you should use are shoulder, elbow, wrist, knuckles, radius, ulna, carpus, metacarpus and phalanges.

Molecular Similarities

Biochemical similarities also demonstrate relationships among various organisms. DNA sequences are studied and compared. The closer the sequences, the more closely related the organisms. Humans and chimpanzees show a great deal of overlap in their DNA sequences. Humans and rabbits show some similarities, but there is less overlap than between the human and chimpanzee sequences. When the DNA from humans and a fruit fly are compared, there is very little overlap. This suggests that humans and chimpanzees are much more closely related than humans and a fruit fly. Figure 8.2 shows a hypothetical comparison of chromosome #7 from different organisms.

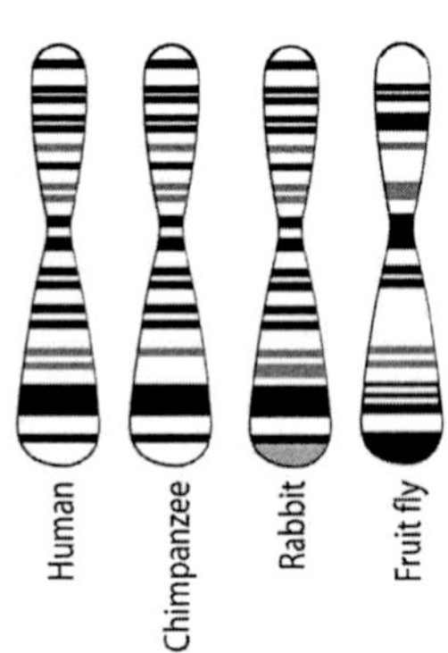

Figure 8.2 Hypothetical Comparison of Chromosome #7 in Animal Kingdom

Embryonic Developmental Similarities

Scientists agree that studying the embryonic development of an organism often leads to a greater understanding of the evolutionary history of that organism. The early development of an embryo is the most important time during its life cycle. Like laying a foundation for a house, the structures and tissues formed at the beginning of development form the basis for many other tissues later in life. Vertebrates (organisms with a backbone or spine) pass through some stages that are similar to each other. The more closely related an organism is, the more similar its stages of development will be. Figure 8.3 shows the embryonic development of four vertebrate animals. Although they spend a different amount of time in each state, they still move through similar stages where they develop comparable structures. One example is an internal skeleton. Initially, all vertebrate animals develop the same skeletal tissues during embryonic development.

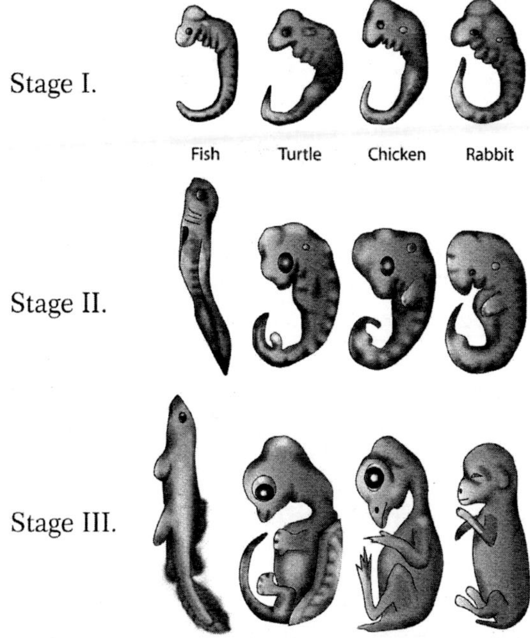

Figure 8.3 Examples of Embryonic Development

THE FOSSIL RECORD

Fossils provide perhaps the most compelling evidence for the change in organisms over time. A **fossil** is the recognizable remains or body impressions of an organism that lived in the past. The study of fossils gives us a fascinating historical perspective — snapshots from an Earth of long ago. Taken together, these snapshots are referred to as the **fossil record**. Scientists use the body of evidence accumulated from the fossil record to make hypotheses about the evolution of organisms. As we will soon see, scientists often see sequential changes in a group of organisms over time. Sometimes they observe maintenance of certain characteristics or organisms over a long period of time. Ferns, sharks and crocodiles are some organisms that have maintained their characteristics over many millions of years.

Samples of iron ore containing ancient volcanic deposits offer clues to the Earth's early atmosphere. These samples lack rust, therefore scientists speculate that our early atmosphere contained very little oxygen. This is known as a reducing atmosphere, which is made up mainly of volcanic gases (methane, carbon monoxide, carbon dioxide, ammonia and water vapor). Two important scientists, Stanley Miller and Sidney Fox, demonstrated that under these early reducing (oxygen-poor) conditions, organic molecules and cell membranes were able to form. Coupling their experimental evidence with ancient bacterial fossils, scientists postulate that anaerobic bacteria were among the first life forms on Earth. Subsequent changes to their cellular structure eventually led to the development of photosynthetic and eukaryotic organisms. Over time, oxygen from photosynthesis eventually filled the atmosphere creating our modern oxidizing atmosphere.

In another more modern case, the evolution of the modern horse is shown through fossils collected from the Eocene epoch (55 million years ago) through modern rock strata. The fossils collected show the clear changes in body structure of the modern horse. Its oldest ancestor, *hyracotherium*, was a forest-dwelling organism about the size of a fox. It had an omnivorous diet consisting of foliage, fruits, flowers and insects. It had many toes on its soft padded feet and short legs. These were adaptations for walking on the soft forest floor. It also had low-crowned teeth filling the entire mouth capable of chewing a variety of foods.

Over time, the lush forest environment changed to become drier with fewer trees and more grasses. In order to survive in the changing prairie-like grassland, *hyracotherium* needed different traits. From Figure 8.4, we can clearly see that over the next few million years the height and size of the horse increased. Notice the sequential nature of the fossils in this group. Its legs lengthened and reduced the number of toes on its feet. Additionally, its teeth became tougher and more capable of grinding grasses.

Eventually the modern horse, *Equus*, developed long legs and tough hooves. These adaptations allowed it to outrun predators. In addition, the larger height allowed *Equus* to see over tall grasses, spotting potential predators before they got too close. Another anatomical change was in the teeth. We can see the modern horse has high-crowned teeth adapted for a lifetime of grinding grassy stalks.

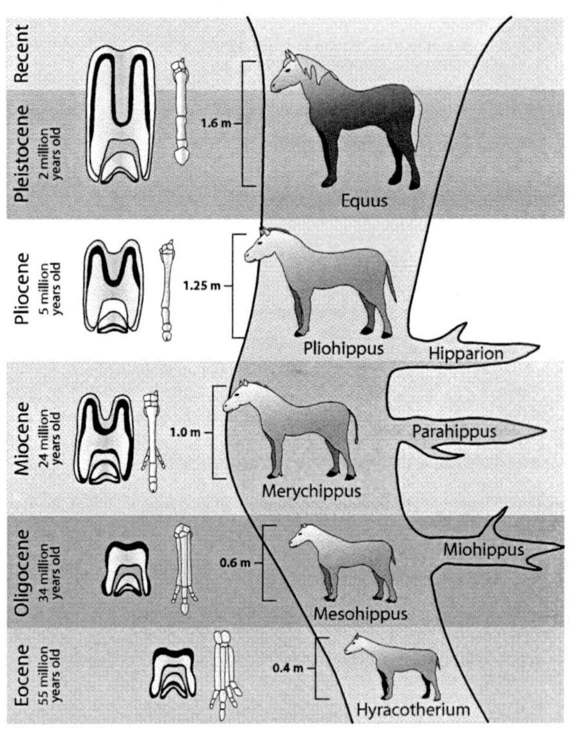

Figure 8.4 Evolution of the Horse

There are plenty of other examples of fossils demonstrating changes over time. Archaeopteryx is one transitional fossil form. This fossil was one of the first to suggest surviving dinosaurs developed homeothermic mechanisms like feathers or fur. Even though it is still somewhat incomplete, the fossil record offers strong evidence demonstrating how organisms can change.

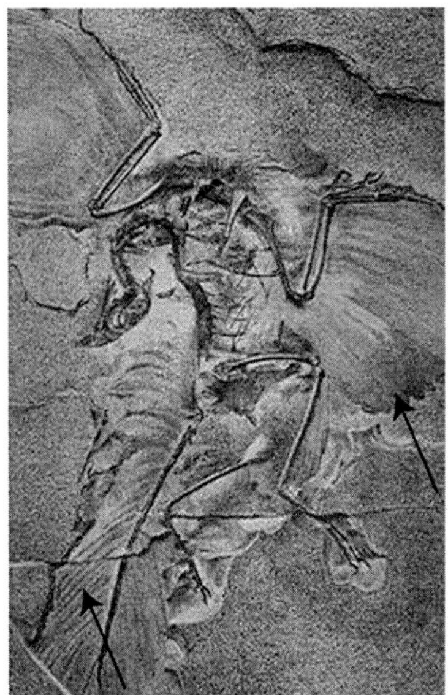

Figure 8.5 Archaeopteryx

Taken together, these clues offer strong evidence for the theory of evolution. Similarities in anatomy, genetic structure and embryonic development undeniably support the idea of a common ancestor. Additionally the gradual descent seen in the fossil record offers direct evidence in support of Darwin's ideas.

Section Review 1: Evidence of Evolution

A. Define the following terms.

fossil	vestigial structure
fossil record	homologous structure

B. Choose the best answer.

1 Two completely unrelated marine organisms have fins. What term **best** describes this situation?

A analogous
B homologous
C vestigial
D coincidence

2 What modern technology assists scientists in determining evolutionary relationships among organisms?

A DNA
B X-ray
C fossil record
D crossbreeding

3 What is **not** used as evidence to support evolution?

A fossil record
B internal skeleton
C genetic similarity
D embryonic development

4 A particular species of cave-dwelling fish developed eye sockets and tiny, nonfunctional "eye bulbs." What word **best** describes this situation?

A homologous
B vestigial
C analogous
D fossil

5 During Stage I of development, a human embryo, a bat embryo and a chicken embryo look strikingly similar. What theory does this support?

A natural selection
B coevolution
C evolution
D fossil record

C. Complete the following exercises.

1 Name and briefly describe the evidence used to support evolutionary theories.

2 Make a concept map or a flip chart showing the evidence for evolution.

IDEAS ON EVOLUTION

Mechanisms of evolution deal with how evolution occurs. Most scientists consider natural selection one of the most important mechanisms of evolution, but other mechanisms like mutations, gene flow and genetic drift are also significant.

NATURAL SELECTION

Naturalist Charles Darwin proposed the idea of natural selection in 1859. Natural selection states that organisms best suited to the environment are the ones most likely to survive and reproduce. A few important points Darwin made in his book, *On the Origin of Species*, are:

- **Resources are limited in all environments**. The availability of food, water and shelter in an environment is limited. This leads to competition among organisms. **Competition** is the fight among living things to get what they need for survival. For example, moths must find food before other moths take all the food.

Figure 8.6 Tree Seeds

- **Most organisms have more offspring than the environment can support**. For example, a fish lays thousands of eggs, or one tree produces millions of seeds in a single season.
- **There is natural variation within a population**. A **variation** is a difference in a trait between organisms within a population. Not all organisms are exactly alike. For example, not all moths are the exact same color. Another example of variation is that not all humans are the same height.

Figure 8.7 Natural Variation in Human Height

- **Natural selection is always taking place**. Organisms with traits that are the most desirable are selected to survive. Organisms in any environment have a specific fitness for that environment. **Fitness** is the ability of an organism to live, survive and reproduce in that environment. Not all of the individual animals within a population have the same fitness. Some organisms are more fit than others — these fit individuals are able to survive the longest, producing the most offspring, thereby, increasing the chances that their offspring will survive to make up a larger portion of the next generation. Darwin's theory of natural selection is also sometimes called "survival of the fittest."

Figure 8.8 Cheetah

Variations in physical characteristics make some organisms better suited, or more fit, to live in their environments. Much of this variation is inherited. For example, the fastest cheetah is better equipped to hunt than a slower cheetah. As a result, the faster cheetah will get more food. The most successful cheetah lives the longest and is, therefore, able to produce the most offspring. Scientists would say that the fastest cheetahs are the most fit for their environment. This is where the idea of survival of the fittest comes from.

Inherited traits that are more versatile than others improve the chances of organism survival and reproduction. For example, having long legs and a skinny body helps a cheetah stay cool. These adaptations also help the cheetah to run fast. These traits are favorable for more than one reason, making them highly likely to be passed on to offspring.

An improved chance of survival allows organisms to produce offspring that will make up more of the next generation, while passing along their favorable traits. Unfavorable traits will eventually be lost, since there is less reproduction among the individuals with such traits. The slowest cheetahs will get the least amount of food and will, therefore, have a greater chance of dying. These slow cheetahs are said to be less fit, or unfit, for their environments. Notice that natural selection acts on the individual members of the population, but overall changes are seen in the population as a whole. Environmental pressures lead to new adaptations that quickly integrate into the larger population. Eventually, the adaptations become a common adaptation that is seen throughout the larger population.

In many cases, the most successful traits are maintained within the population and change very little over a long period of time. Remember the sharks, crocodiles and ferns mentioned earlier?

Natural Selection and Diversity

Environmental conditions contribute to variations in traits among individuals of the same species. This increases the diversity among members of the same population. The size of house sparrows in North America varies depending on location. House sparrows living in colder climates are larger than those living in warmer climates. As a general rule, the larger the body size of an animal, the more body heat it can trap or conserve.

The size of extremities, such as ears or legs in some animals, also demonstrates environmental differences. Since extremities give off heat to help cool the animals' bodies, mammals living in hot climates tend to have larger ears and longer legs than their cousins in cooler climates. A desert jackrabbit has much larger ears than a rabbit found in a temperate (cooler winter) climate.

Figure 8.9 Desert and Temperate Rabbits

Natural variation within a population allows for some individuals to survive over other individuals in a changing environment. Recall our earlier example in the fossil record section: the evolution of the modern horse. The changing environment selected individuals within the population that had the longest legs, toughest teeth and could run fast. The natural variation within a population can eventually lead to the formation of new species, which is called **speciation**. Speciation creates a greater diversity among organisms. Speciation is constantly happening. Organisms with fast generation times — ones that produce many offspring several times a year — can quickly create many different species. Organisms with slow generation times tend to take a lot longer to create new species. Think about this: Flies live for 30 days or less, whereas elephants can live for decades — as a result, there are over 120,000 known species of flies and only 3 living species of elephant.

Mutations

Mutations are random changes in DNA that act as further mechanisms for evolution. Mutations in DNA can come from disease agents like viruses, bacteria or environmental chemicals. These changes result in a variation in traits, which then are passed on from one generation to the next. Mutations can be beneficial, neutral or harmful to an organism. Mutations beneficial to the organism in a particular environment lead to furthering of the species. For example, a mutation can result in the production of an enzyme that breaks down a particular food product predominant in an area. Individuals with the expression of that gene have more food choices, giving them greater survival chances and allowing them to be more successful. Another example could be a mutation in color pigments that leads to an individual that is a different color than the normal population. This might allow better camouflage, thereby contributing to a longer lifespan and an increased number of offspring.

Gene Flow

Gene flow is the exchange of genes between two populations. It most often occurs due to migration or movement of the organisms. Gene flow occurs when individuals leave a population or new individuals join a population. When individuals join a population, their physical and genetic similarities increase.

When individuals leave a population, their genetic destiny can become quite different. Individuals that depart from a population take their genes with them, effectively shrinking the gene pool of the larger population and isolating themselves. This is another way speciation can happen. Sometimes geographic isolation can quickly lead to speciation. Darwin's finches are perhaps the most famous example of geographic isolation leading to the formation of a new species. On the Galapagos Islands (an isolated environment) all the bird species on the islands are a unique type of modified finch. These birds closely resemble one species of South American finch, yet they are distinctly different. There are over a dozen different species of Galapagos finch. Due to the geographic isolation of the island and consequent genetic restrictions over many generations, one bird species was able to develop into more than a dozen different types.

Figure 8.10 Individuals Join a Population

Figure 8.11 Individuals Leave a Population

Genetic Drift

Figure 8.12 Bottleneck

Genetic drift is the change in frequency of alleles in a population. It provides random changes in the occurrence of genes through chance events. Small populations are more affected by genetic drift than larger populations. **Bottlenecking** can occur if a large number of the population is killed because of disease, starvation, change in natural environment or a natural disaster. When this happens to a population a large population is reduced to a few individuals, and the genes of subsequent generations become very similar. Bottlenecking can lead to the quick development of new species, or it can contribute to the extinction of a species.

Inbreeding between these few individuals leads to populations that have very few genetic differences. It is believed that African cheetahs went through two genetic bottlenecks, one about 10,000 years ago and one about 100 years ago. All African cheetahs alive today are descendents of a few cheetahs — and possibly only three females. Because cheetahs are genetically similar, they have become very susceptible to diseases.

Patterns of Evolution

The theory of evolution suggests that there is more than one way to evolve or change. These different patterns provide different paths to explain the degree of variation among organisms. Some ways that organisms evolve, or change, include: convergent evolution, divergent evolution and coevolution.

Convergent Evolution

Convergent evolution explains how unrelated species can develop similar characteristics. Convergent evolution is demonstrated through the porpoise and the shark. The porpoise is a mammal, and the shark is a fish. These two unrelated animals share similar characteristics that suit their environment: long, streamlined bodies and fins that closely match in both appearance and function. These structures are said to be analogous. **Analogous** structures are similar in function but have different ancestors. Another example of analogous structures included the wing of a butterfly and the wing of a bird.

Figure 8.13 Convergent Evolution of Porpoise and Shark

Convergent evolution also occurs in the plant kingdom. Cacti and specimens of *Eupherbia* are both plants that look very similar and live in desert climates. They both have spines, small leaves and water storage tissues in large, fleshy stems. Cacti are found in North America, and euphorbs are found in Asia and Africa. Despite the similarity in characteristics, these plants have very different flowers and are not closely related. These organisms have evolved similar characteristics to suit their specific environments. Their spines are analogous structures, not homologous structures.

Divergent Evolution

Divergent evolution suggests that many species have developed from a common ancestor. These different species change as they adapt to their particular environments. A good example of divergent evolution is Darwin's finches. A few birds from one species of finch were carried to the Galapagos Islands. Since their food supply changed in the new locations, they adapted their diets accordingly. Geographic isolation from other finches and environmental differences caused them to diverge from their ancestors and even from one another. There are now many recognized species of finches, all diverging from a common ancestor. Divergent evolution is demonstrated by homologous structures.

COEVOLUTION

Coevolution occurs when two or more organisms in an ecosystem evolve in response to each other. Coevolution is believed to occur frequently with flowers and their pollinators.

Historical and modern botanists alike study the prolific orchid family for clues to evolutionary processes. Darwin himself became interested in orchid diversity and their methods of pollination.

Orchids have close relationships with their pollinators, making them good examples of coevolution. In perhaps one of the most extreme examples, the Christmas orchid or Comet orchid (*Angraeeum sesquipedale*) has a tube 20 – 35 cm (1½ ft.) long where it keeps nectar. In 1862, Darwin hypothesized that moths pollinating this flower must have a tongue (proboscis) reaching at least 25 – 30 cm (10 – 12 in.). Although he did not know it to be true, Darwin speculated that the moth *must* exist, given the length of the nectar tube. He was ridiculed and criticized by many biologists for this seemingly absurd speculation.

Figure 8.14 Comet Orchids

Figure 8.15 Morgan's Madagascan Sphinx Moth

Nearly 40 years later, after Darwin's death, in 1903 the Morgan's Madagascan sphinx moth (*Xanthopan morgani*) was discovered. This hawk moth has a proboscis 30 – 35 cm (12 – 14 in.) long. In this example of coevolution, as the orchid lengthened its nectar tube (to ensure proper pollination), the moths with the longest tongues were rewarded with nectar. Thus moths with long tongues survived to create offspring with long tongues. Additionally, pollinated orchids with long nectar tubes created seeds for offspring flowers with long nectar tubes.

HUMANS AND NATURAL SELECTION

For thousands of years, disease and starvation, among other factors, limited the human population. Like all other living creatures, humans produce far more offspring than the environment can support. During the Industrial Revolution, the human population experienced explosive population growth that continues to this day. This growth is due, in part, to advancements in medical technology and agricultural sciences. Vaccines and antibiotics are two medical technologies that have allowed humans to slow the impact of disease on the population. Chemicals like pesticides have initially improved crop yields, decreasing starvation's impact on the population.

It all began in 1796 when Edward Jenner developed the first vaccine. A **vaccine** is a medical solution that contains a weakened form of the disease agent like a virus. In Jenner's case, he noticed that milkmaids who were first exposed to the cowpox virus seemed unaffected by subsequent exposure to the more deadly smallpox virus. His experiments led to the development of a smallpox vaccine. Vaccines work by activating the body's natural immunity. Figure 8.16 summarizes this process.

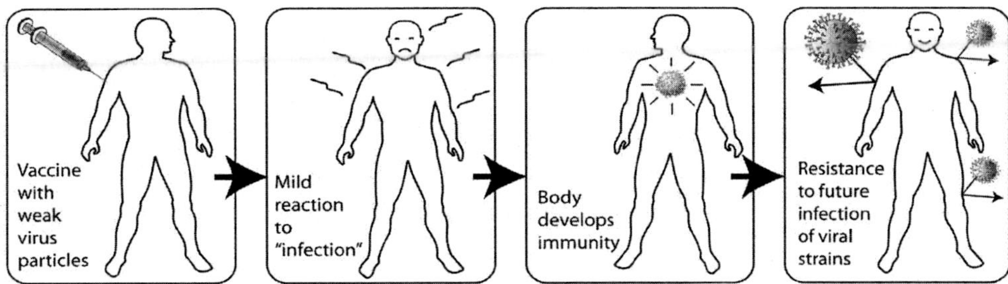

Figure 8.16 How a Vaccine Works

Immunity is the ability to resist disease. **Active immunity** is acquired when an immune response occurs in the human body.

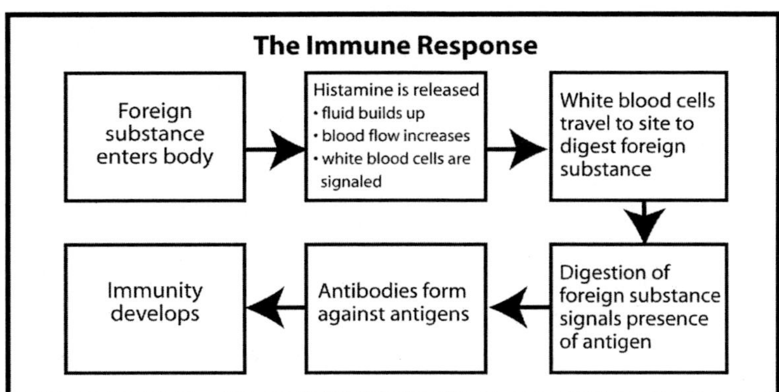

Figure 8.17 The Immune Response

The immune response is the body's way of removing harmful disease-causing pathogens like viruses and bacteria. Active immunity is long term and can provide a lifetime of disease protection. This is how Jenner's observations led to his smallpox vaccine development. Milkmaids who were exposed to a weakened cowpox virus developed an active immune response to all pox viruses. This made them immune to the smallpox infection.

Another type of immunity is passive immunity. In **Passive immunity**, antibodies are introduced into the body. Passive immunity often occurs when mothers breastfeed infants. Antibodies present in the mother's milk passes to the infant during feedings. This type of immunity is short term. It is useful to infants during the first year of life while their immune system matures.

Using the body's natural disease-fighting agents is a powerful tool against sickness. In 1979, the World Health Organization certified that smallpox has been eliminated from the human population. This achievement was the culmination of vigorous vaccination programs throughout the world. Other diseases which have been controlled through vaccination programs include polio, measles, mumps, rubella, whooping cough, chickenpox and tetanus. Vaccination programs have greatly reduced the selective impact of disease on the human population.

Antibiotics are another tool used by humans to prevent disease. Antibiotics are chemical compounds that prevent microbial growth. Usually used to treat bacterial infections, antibiotics are often derived from funguses or plant materials. This medical technology has saved thousands of lives over the past fifty years.

Irresponsible use of this technology has created several dangerous strains of **antibiotic resistant bacteria.** Over the years, bacteria have developed resistance to the chemicals used in antibiotics. This is simply natural selection at work. Here's how it happened. When someone had an infection, and they didn't complete the doctor's recommended course of antibiotics, a certain percentage of the bacteria that caused the infection survived. These bacterial survivors had offspring that also showed resistance to antibiotics. Figure 8.18 shows how a patient can create an antibiotic resistant bacteria. This patient visits a doctor and receives a 14 day antibiotic prescription. Notice that the patient stops taking the medication before the 14 day dosage is up. The surviving bacteria continue growing and creating offspring resistant to the antibiotic.

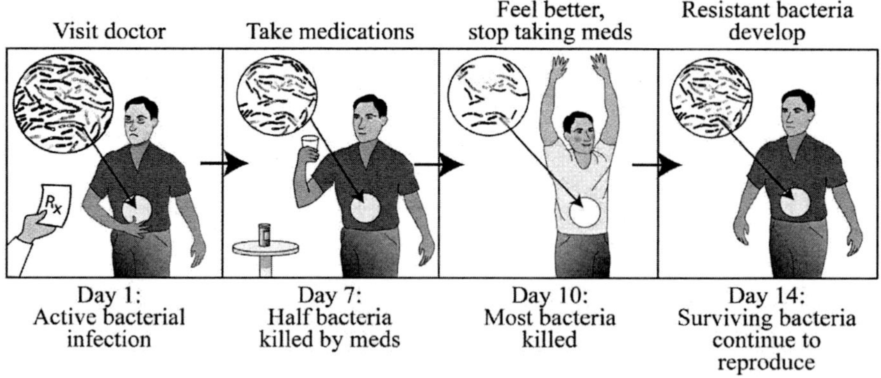

Figure 8.18 How an Antibiotic Resistant Strain of Bacteria is Created

A similar situation has occurred in agricultural technology. Modern chemicals like pesticides have greatly reduced the impact of pests on crops. However, pesticides are chemicals used to control insect populations. When a pesticide is applied to an area, a certain percentage (usually 95 – 99%) of the target organism is killed, along with other non-target organisms. Only a small fraction of the original population is left intact. However, this small surviving fraction of the population passes along its immunity to future generations. Let's say, for example, a tomato farmer has a problem with snails. He estimates his field has a population of about 1,000,000 snails. He applies metaldehyde, a type of mollusk pesticide, to his field. This chemical is 97.1% effective, and kills 971,000 of the snails on his field. The remaining 29,000 snails have a resistance to the pesticide. These snails reproduce, passing along their resistance to the next generation. Examine Figure 8.19, showing five hypothetical years on the tomato farm.

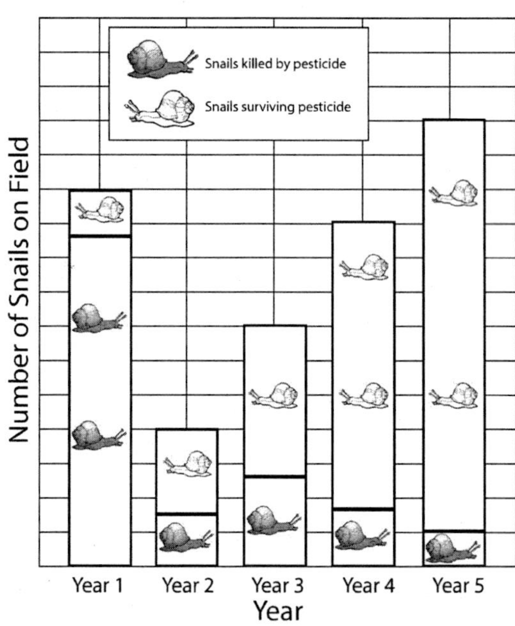

Figure 8.19 Snails in the Tomato Garden

Pesticide resistance and antibiotic resistance are two ways humans have impacted natural selection. In these two examples, our technological ability has outstripped our evolutionary common sense.

Section Review 2: Mechanisms and Patterns of Evolution

A. Define the following terms.

speciation	gene flow	bottlenecking	coevolution
competition	genetic drift	convergent evolution	genes
variation	analogous structures	divergent evolution	mutations
fitness			survival of the fittest

B. Choose the best answer.

1 What are the effects of genetic drift and gene flow?

A change in gene occurrences
B change in vision acuity
C change in DNA replication patterns
D change in organism size

2 Which of the following are patterns of evolution?

A structural replication, reproductive homology and special creation
B metabolic pathways, hormonal indicators and genetic studies
C modern creationism, fossil theory and punctuational models
D convergent evolution, coevolution and divergent evolution

3 If two organisms evolve in response to each other, which evolutionary pattern is demonstrated?

A divergent evolution
B emigration
C coevolution
D convergent evolution

4 Darwin identified more than a dozen different species of finch during his time on the Galapagos Islands. The main difference between the finches was the size and shape of their beaks. Which of the following statements **best** describes how these differences developed?

A bottlenecking
B coevolution
C convergent evolution
D divergent evolution

C. Complete the following exercises.

1 What is the difference between convergent and divergent evolution? Give an example of each.

2 Explain the theory of natural selection.

3 How is variation beneficial to an organism? How is it harmful?

4 Explain antibiotic resistance.

Chapter 8 Review

1 Identify the evidence for evolution.

A cave drawings, ancient stories and ceremonial rites

B homologous structures, DNA and embryonic evidence

C eukaryotes, symbiosis and competition

D nephrons, antibodies and homeostasis

2 A mountain, ocean or ravine divides a population. After many years, the organisms show genetic differences from the original population. Which of the following explains how this change occurred?

A divergent evolution

B convergent evolution

C coevolution

D immigration

3 Humans have an appendix, a thin tube connected to the large intestine that serves no purpose and is a threat to human health and life if it becomes infected and/or inflamed. It is believed that the appendix once had a function as part of the human digestive system. The human appendix, therefore, is

A a homologous structure.

B a vestigial organ.

C a vital organ.

D a mutation.

4 In order to find the best nesting sites and fewest predators, a group of Arctic gulls moves farther and farther north. The birds only begin breeding once they reach their isolated nesting grounds. Over time, how would you expect their appearance to change with respect to the original gull population?

A They will look the same with no change.

B They will look similar with little change.

C They will look different with a few changes.

D They will look completely different and be unrecognizable.

5 Sharks and porpoises are a classic example of which concept?

A coevolution

B convergent evolution

C divergent evolution

D parallel evolution

6 Increased use of antibiotics has killed off bacterial populations that were most susceptible to antibiotic treatment. Consequently, many strains of bacteria are resistant to prescription drugs. What is the mechanism by which these resistant bacteria have been allowed to thrive?

A natural selection

B mutation

C speciation

D germination

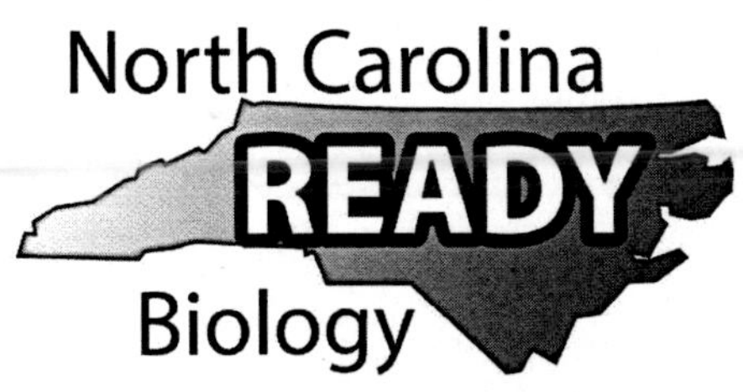

Chapter 9 Classification

This chapter covers:
3.5.1, 3.5.2

BIOLOGICAL CLASSIFICATION

.Biologists classify living things according to the traits they share. **Taxonomy** is the classification of an organism based on several key features such as structure, behavior, lifecycle, genetic makeup (DNA), nutritional needs and methods of obtaining food. Evolutionary theory is the basis for this classification system. Taxonomy divides organisms into several categories that start out broadly and become more specific. These categories are **kingdom**, **phylum**, **class**, **order**, **family**, **genus** and **species**.

Occasionally, subphylum, subclasses and suborders are used to further delineate characteristics among the primary classifications.

To remember the order of the subdivisions, memorize the silly sentence, "King Phillip Came Over From Greece Sneezing." The first letter of each of the words in this sentence is also the first letter of each of the classification categories for organisms.

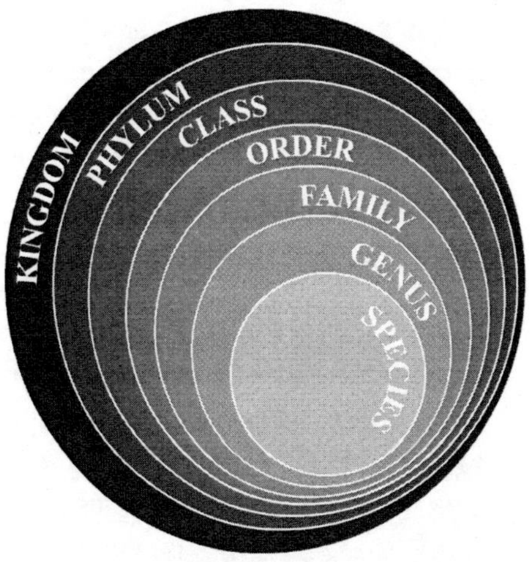

Figure 9.1 Classification System for Organisms

Table 9.1 lists the six **kingdoms** based on general characteristics. Each kingdom further divides into **phylum**, to name organisms in the kingdoms of Eubacteria. Phylum further break down into **classes**, and classes break down into **orders**. The categories become progressively more detailed and include fewer organisms as they are further broken down into **family**, **genus** and **species**. The species is the most specific category. Organisms of the same species are grouped together based on their ability to breed and produce fertile offspring.

Classification

Table 9.1 The Six Kingdoms

Super Kingdom	Kingdom	Basic Characteristic	Example
Bacteria	Eubacteria	found everywhere	cyanobacteria
Archaea	Archaea	live without oxygen, gets its energy from inorganic matter or light, found in many habitats	halophiles
Eukaryota	Protista	one-celled or multicellular, true nucleus	amoeba
	Fungi	multicellular, food from dead organisms, cannot move	mushroom
	Plantae	multicellular, cannot move, make its own food, cell walls	tree
	Animalia	multicellular, moves about, depends on others for food	horse

Aristotle (384 – 322 BC) made the first recorded attempt at classification of plants and animals. He grouped all living things into two main categories: plants and animals. According to Aristotle, animals were further divided into two groups: blooded and bloodless.

Aristotle's classification system had many flaws because the relatedness and reproductive strategies of organisms was not completely understood in his time. However, Aristotle's contribution to taxonomy was important because early on it promoted the scientific way of organizing information.

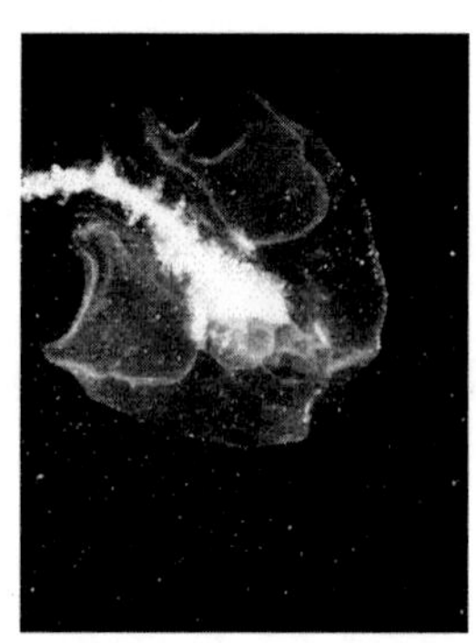

Figure 9.2 A "Bloodless" Animal

Carl Linnaeus (1707 – 1778), a Swedish botanist, devised the current system for classifying organisms. Linnaeus used **binomial nomenclature**, a system of naming organisms using a two-part name, to label the species. The binomial name is written in Latin and is considered the scientific name. It consists of the generic name (genus) and the specific epithet (species). The entire scientific name is italicized or underlined, and the genus name is capitalized, as in *Homo sapiens* for humans. Table 9.2 is a complete classification of three members of the kingdom Animalia.

Figure 9.3 Carl Linnaeus

A classification system is necessary to distinguish among the great number of organisms and to avoid confusion created by the use of common names. Common names are used for many organisms, but not all organisms have common names, and some have multiple common names.

Table 9.2 Examples of Classifications

Example:	Human	Grasshopper	Dog
Kingdom	Animalia	Animalia	Animalia
Phylum	Chordata	Arthropoda	Chordata
Class	Mammalia	Insecta	Mammalia
Order	Primate	Orthoptera	Carnivora
Family	Homindae	Locuslidea	Canidae
Genus	*Homo*	*Schistocerca*	*Canis*
Species	*sapiens*	*americana*	*familiaris*

The hierarchical classification devised by Linnaeus has been, and still is, quite useful in organizing organisms. However, limitations do exist. For instance, even though classification is based on evolutionary theory, it does not reflect the idea that evolutionary processes are continual, and species are not fixed. Remember the term speciation, which denotes a continual process. Changes to organisms will occur over time and, therefore, classification will also have to change. Also, classification does not take into account the variation that exists among individuals within a species. All domestic dogs have the scientific name *Canis lupus familiaris*, but a great deal of variation exists among different breeds of dogs and even among individual dogs of the same breed. Think about the difference between a Great Dane and a Chihuahua.

Figure 9.4 Great Dane vs. Chihuahua

Finally, the most definitive test to determine if organisms are of the same species is to confirm their ability to breed successfully, producing fertile offspring. However, controlled breeding of wild organisms for the purpose of observation and study can sometimes be impractical, if not impossible. Also, sometimes closely related species can interbreed, such as in the mating of a horse and donkey to produce a mule. Classification has been instrumental in bringing about an understanding of similarities and possible evolutionary relationships of organisms. However, it is not static and may need to change with the discovery of new organisms and as more evidence of evolutionary patterns surface.

THE THREE DOMAINS

Figure 9.5 shows the current three-domain classification system. Notice the "branching tree" appearance. Think of all the different types of organisms found in the world. From a single-celled bacterium to a multicellular human, organisms can have a great diversity of structures. In the next section, we will examine ways to compare organisms throughout history.

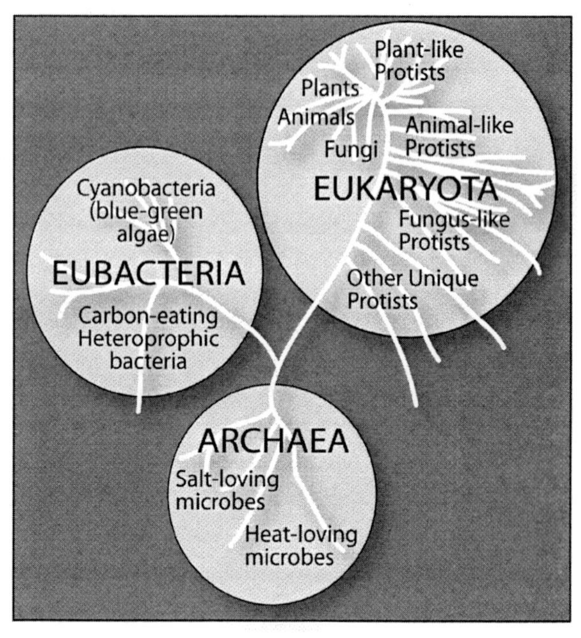

Figure 9.5 The Three-Domain Classification System

Section Review 1: Classification

A. Define the following terms:

Carl Linnaeus	phylum	genus
taxonomy	class	species
kingdom	order	Aristotle
	family	binomial nomenclature

B. Choose the best answer.

1 Which category of taxonomy is ***most*** inclusive?

A phylum
C kingdom
B genus
D species

2 Which category of taxonomy is ***most*** restrictive?

A phylum
C kingdom
B genus
D species

3 Which domain contains vertebrate animals?

A Archaea
C Eukaryota
B Eubacteria
D Nomenclature

PHYLOGENETIC TREES

Phylogeny is the study of the relatedness among various organisms. The evolutionary idea that all organisms are descended from a common ancestor forms the basis of phylogeny. This is at the very heart of our understanding of biology. Genetic information provides evidence for the phylogeny of organisms. The phylogeny of a particular organism shows how its lineages have changed through history. Phylogenic information about organisms is used to create a phylogenetic tree. A **phylogenetic tree** is a biological model that shows relatedness among various organisms. In this model, living organisms form the leaves of the tree. Working down a branch, you discover various ancestral forms of organisms. Branches diverge as new species form. Figure 9.6 shows one example of a phylogenetic tree.

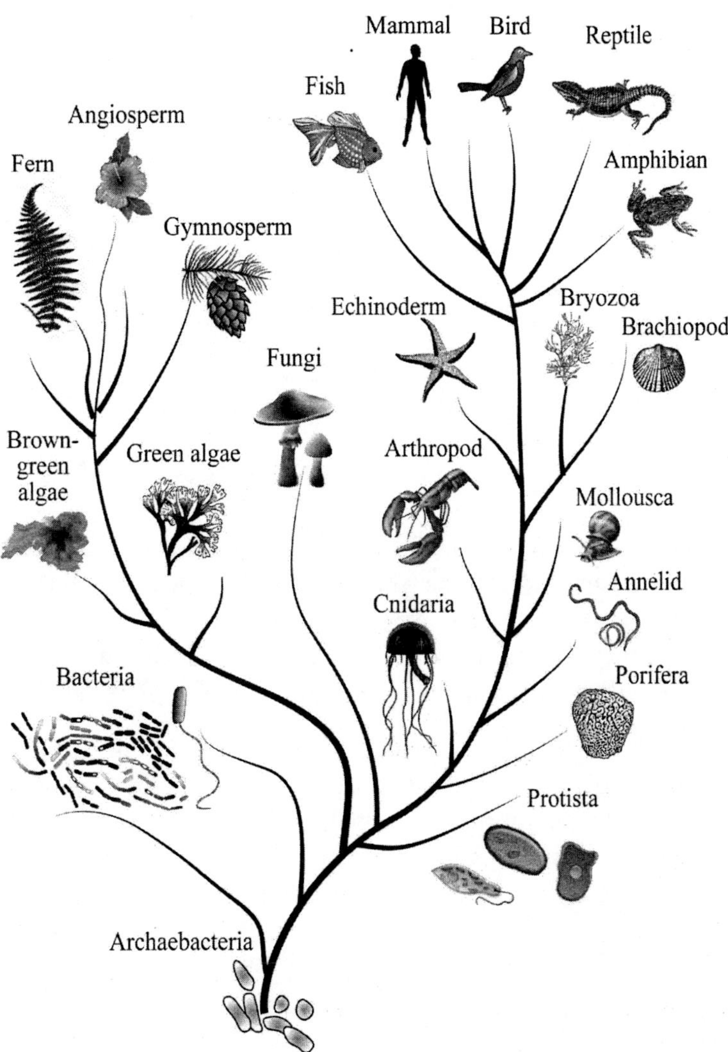

Figure 9.6 The Tree of Life

Following a particular organism's branch backward down the tree, you see its various ancestors from hundreds or even millions of years ago. The length of a line can offer some clues to the amount of time that has passed since the two species diverged. In Figure 9.7 you will notice that Point 1 is further down the tree when compared to Point 3. This means that the divergence of Point 1 occurred much earlier than Point 3. Now look at the divergence between Point 3 and Point 4, we can see this departure occurred around a similar time.

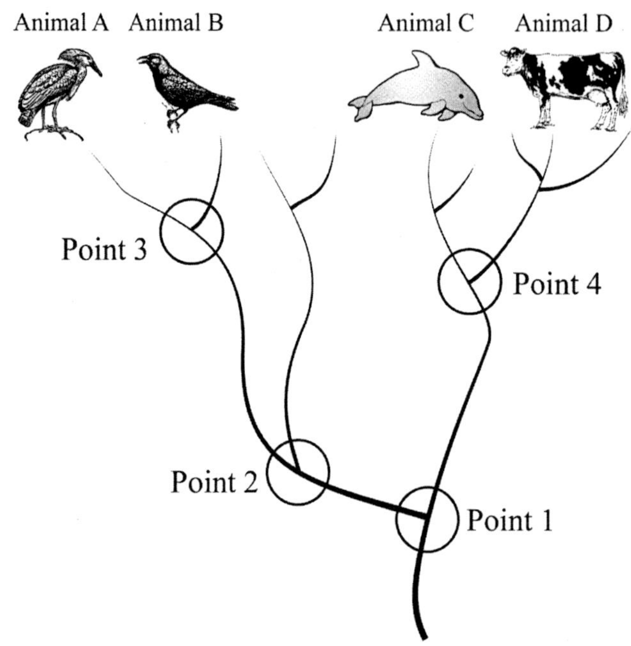

Figure 9.7 The Tree of Life Points

We can look at our simple tree to determine the relatedness of the organisms on it. In Figure 9.7, notice that organisms located on the same branch are very similar. This is because they share a common genetic ancestor. In Figure 9.7, organisms A and B are closely related. Organisms with recent ancestors often share many genetic characteristics. Organisms found on different branches are distantly related. In our figure, organisms A and D are only distantly related. We use the tree to observe that their most recent common ancestor occurred at Point 1.

Phylogenetic trees are a good model to demonstrate ways that organism have diversified over time. Remember it is based on genetic information and shows the most related organisms close together on the same branch. As newer genetic information is gathered and the relationships among the various plant and animal species are better understood, the phylogeny of these groups is adjusted to show the most current information. This tool can quickly and easily help you decide which organisms are most related.

Using a Dichotomous Key

The identification of biological organisms can be performed using tools such as the dichotomous key. A **dichotomous key** is an organized set of questions, each with yes or no answers. The paired answers indicate mutually exclusive characteristics of biological organisms. You simply compare the characteristics of an unknown organism against an appropriate dichotomous key. The key begins with general characteristics and leads to questions which indicate progressively more specific characteristics. By following the key and making the correct choices, you should be able to identify your specimen to the indicated taxonomic level.

An example follows that uses known organisms. Pick an organism, and then use the key below the illustration to determine its taxonomic classification.

1. Does the organism have an exoskeleton?

   Yes... Go to 2.

   No... Go to 4.

2 Does the organism have 8 legs?

   Yes... It is of class Arachnida, order Araneae.

   No... Go to 3.

3 Does the organism dwell exclusively on land?

   Yes... It is of phylum Arthropoda, subphylum Crustacean, class Malacostraca, order Isapoda, suborder Dniscidea.

   No... Go to 4.

4 Does the organism have an endoskeleton?

   Yes... Go to 5.

   No... Go to 6.

5 Does the organism dwell exclusively in the water?

   Yes...Go to 6

   No... Go to 7

6 Does the organism have stinging tentacles?

Yes... It is of phylum Cndaria, class Scyphozoa

No... Go to 7.

7 Does the organism have 5 legs?

a. Yes... It is of phylum Echinodermata, class Asteroidea

b. No... Go to 8.

8 Does the organism carry its live young in a pouch?

a. Yes... Go to 9.

b. No... Go to 10.

9 Does it climb trees?

a. Yes... It is of class Mammalia, subclass Marsupalia, order Diprodonia, suborder Vombatiformes.

b. No... It is of class Mammalia, subclass Marsupalia, order Diprodonia, suborder Phalangerida, genus *Macropus*

10 Is the organism a mammal?

a. Yes.... Go to 11.

b. No... It is of phylum Chordata, class Actinoptergii, order Perciformes, family Scrombridae, genus *Thunnus*

11 Does the adult organism have teeth?

a. Yes... It is of phylum Chordata, class Mammalia, order Cetacea, suborder Odontoceti.

b. No... It is of phylum Chordata, class Mammalia, order Cetacea, suborder Mysticeti.

Were you able to identify all the animals? If not, one glitch might be that some of these subcategories go beyond the knowledge that has been outlined in our text. These are easily investigated by going online and searching simply for the animal name. You will be surprised at how much you learn.

Section Review 2: Organism Relationships

A. Define the following terms.

phylogenetic tree dichotomous key

B. Choose the best answer.

1 Examine the picture below, and then use the key to identify the animal.

Key	
1. Is the animal an insect with wings?	
a. Yes	Go to 2.
b. No	Use another key.
2. Does the insect have hind legs adapted for jumping?	
a. Yes	order Orthoptera
b. No	Go to 3.
3. Does the insect have front legs adapted for catching prey?	
a. Yes	order Mantodea
b. No	Go to 4.
4. Does the insect have a stinger on the abdomen?	
a. Yes	order Hymenoptera
b. No	order Diptera

A Orthoptera B Mantodea C Hymenotera D Diptera

2 Use the key above to identify the insect below?

A Orthoptera

B Mantodea

C Hymenotera

D Diptera

CHAPTER 9 REVIEW

1 Which taxonomic classification below contains the **fewest** organisms?

A kingdom B family C order D class

2 What was Aristotle's contribution to taxonomy?

A He developed the current system of classification still in use today.

B He established the idea of grouping similar living things together.

C He taught others how to identify living things based on their type of reproduction.

D He made no noticeable contribution to the field of taxonomy.

3 Which kingdom listed below contains prokaryotic organisms?

A Archaea B Plantae C Animalia D Fungi

4 What is **not** one of the three domains of living things?

A Eubacteria B Archaea C Angiosperms D Eukaryota

Use the diagram below to answer questions 5 and 6.

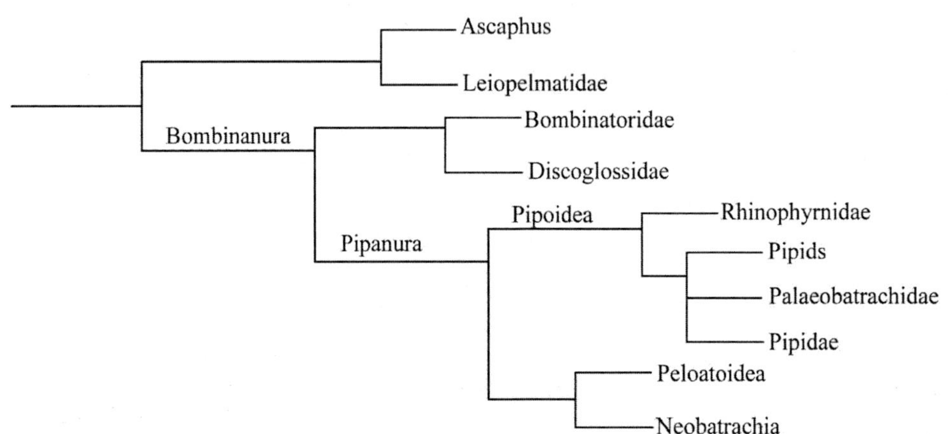

5 Which two organisms are ***most*** related?

A Pipids and Neobatrachia

B Ascaphus and Peloatoidea

C Bombinatoridae and Discoglossidae

D Leiopelmatidae and Pipidae

6 Which two organisms are ***least*** related?

A Ascaphus and Leiopelmatidae

B Peloatoidea and Neobatrachia

C Pipidae and Palaeobatrachidae

D Rhinophrynidae and Ascaphus

7 Look at the image below, and use the key to identify the animal.

Key

1.	a. Tail moves side to side	Go to 2.
	b. Tail moves up and down	Go to 3.
2.	a. Body covered with spots	*Rhincodone typus*
	b. Dark grey body with white underbelly	*Carcharodon carcharisas*
3.	a. Black and white body	*Orcinus orca*
	b. White body	*Delphinapterus leucas*

A *Rhincodone typus*

B *Carcharodon caracharias*

C *Orcinus orca*

D *Delphinapterus leucas*

8 Use the key above to identify the animal pictured below.

A *Rhincodone typus*

B *Carcharodon caracharias*

C *Orcinus orca*

D *Delphinapterus leucas*

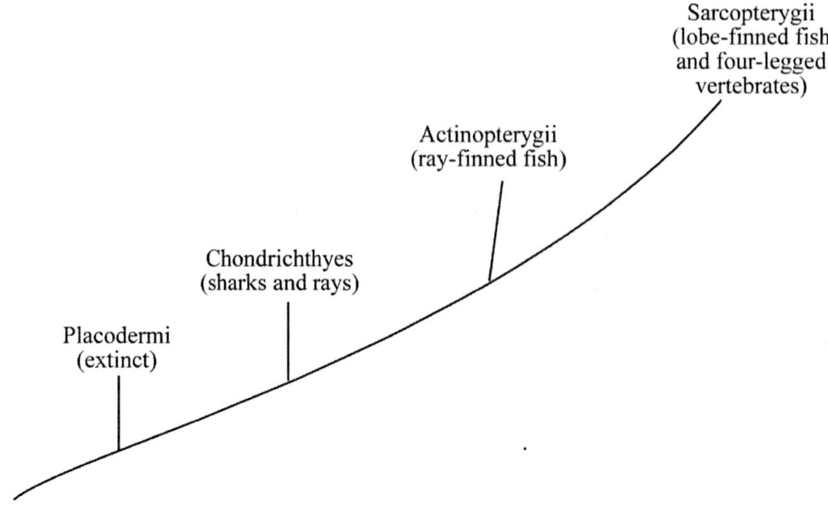

9 Which organism group appeared last in the fossil record?

A Sarcopterygii

B Actinoptergii

C Chondricthyes

D Placodermi

10 Examine the phylogenetic tree below. 3.5.2

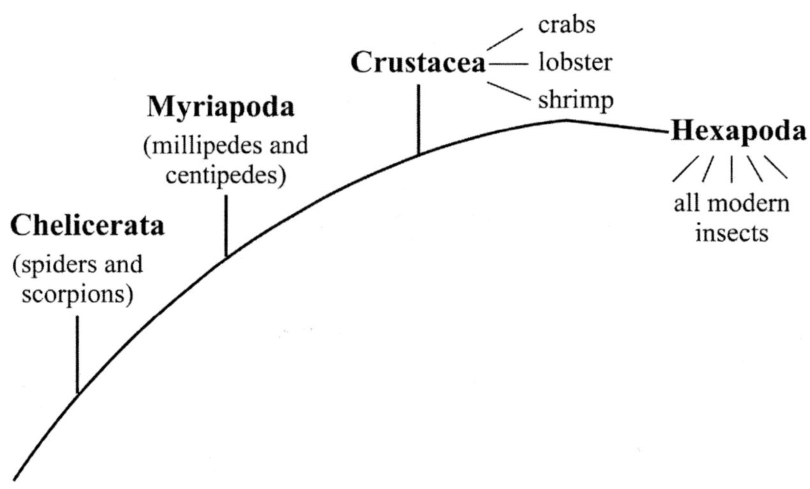

Which group appeared first in geologic history?

A Hexapoda

B Crustacea

C Myriapoda

D Chelicerata

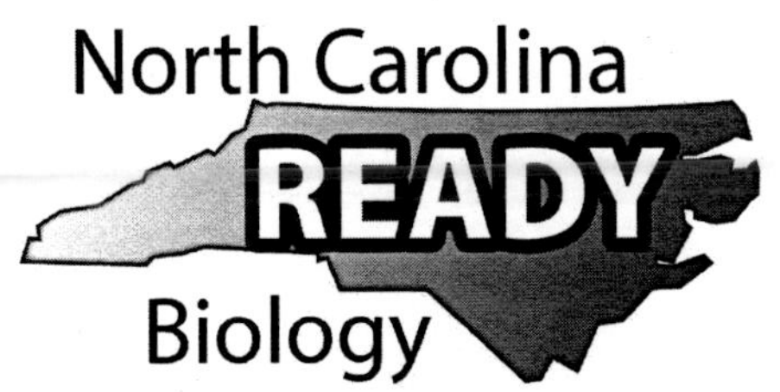

Chapter 10 Biomolecules

This chapter covers:
4.1.1, 4.1.2, 4.1.3

CHEMISTRY OF THE CELL

KEY ELEMENTS

An **element** is a type of matter composed of only one kind of atom which cannot be broken down to a simpler structure. There are six elements commonly found in living cells: **sulfur**, **phosphorus**, **oxygen**, **nitrogen**, **carbon** and **hydrogen** (easily remembered as **SPONCH**). These elements make up 99% of all living tissue and combine to form the molecules that are the basis of cellular function. Carbon is especially important because one carbon atom can make covalent bonds with four other atoms, resulting in the formation of very stable and complex structures. Carbon is present in all living things, as well as in their remains. Molecules containing carbon are called **organic molecules**, while those without carbon are called **inorganic molecules**. Water is the most important inorganic molecule for living things and serves as the medium in which cellular reactions take place.

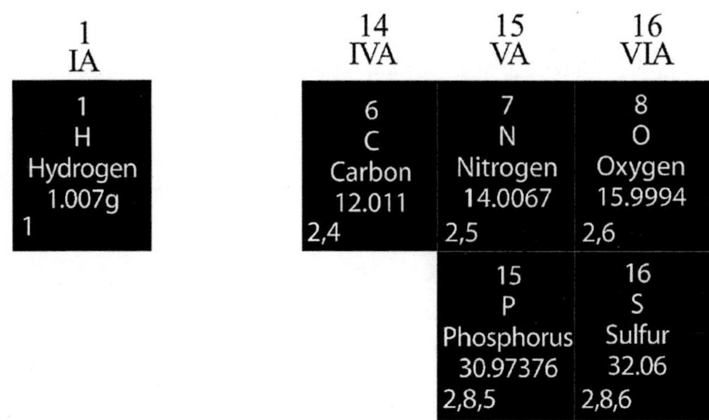

Figure 10.1 Key Elements in Living Cells

Those cellular reactions occur in great part between biological molecules, often called **biomolecules**. The four primary classes of cellular biomolecules are carbohydrates, lipids, proteins and nucleic acids. Each of these is a **polymer** — that is, a long chain of small repeating units called **monomers**.

CARBOHYDRATES

Carbohydrates are often called sugars and are an energy source. Structurally, they are chains of carbon units with hydroxyl groups (-OH) attached. The simplest carbohydrates are **monosaccharides**. The ends of these sugars bond and unbond continuously, so that the straight-chain and cyclic (ringlike) forms are in equilibrium. Figure 10.2 shows a Fischer diagram projection of glucose, a very common biomolecule. A Fischer projection depicts the straight chain form of a monosaccharide. Figure 10.3 shows a Hayworth representation of **ribose**, another common carbohydrate. A Hayworth representation indicates the structure of a cyclic monosaccharide.

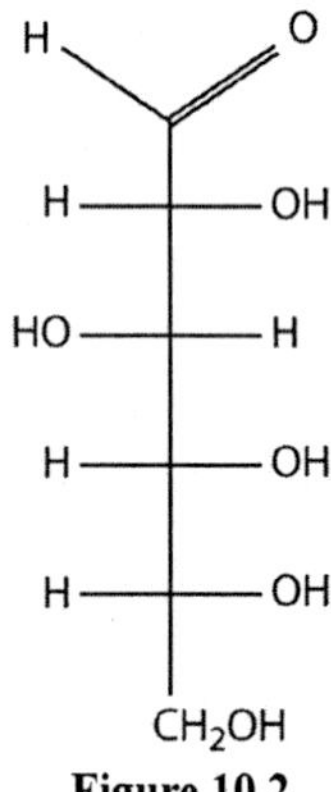

Figure 10.2
Fischer Diagram of Glucose

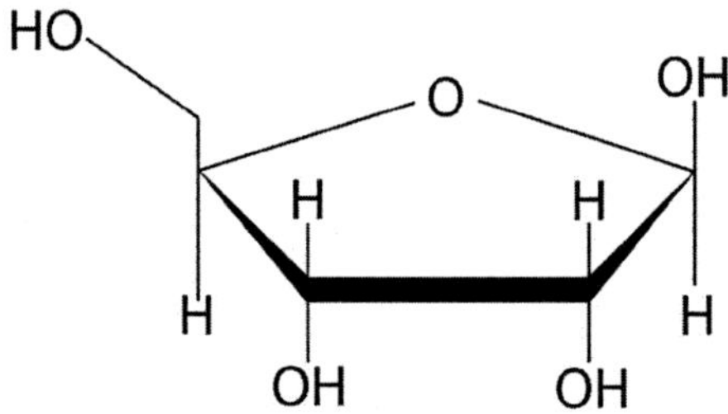

Figure 10.3 Hayworth Diagram of Ribose

These monosaccharides may join together to form **disaccharides** (2), **oligosaccharides** (3 – 10) or **polysaccharides** (10+), depending on how many monosaccharides make up the polymeric carbohydrate. Disaccharides consist of two monosaccharide units. Common table sugar, or sucrose, is a disaccharide formed from the bound monosaccharides, fructose and glucose. Oligosaccharides are made up of 3 –10 monosaccharide units. Oligosaccharides are sugars that are either being assembled or broken down, so there aren't any well-known common names for them. Polysaccharides consist of ten or more monosaccharide units. Complex carbohydrates such as starch and cellulose are classified as polysaccharides.

Typically, carbohydrates serve as the main energy source for the body. In biomolecules, their energy is stored in the chemical bonds of the molecule. A large complex carbohydrate has many carbon-hydrogen bonds and therefore has a high calorie number. A **calorie** is the amount of energy stored in a molecule. It is often used to give people a general idea regarding the amount of energy in food. Chemists define a food calorie as the amount of heat energy required to raise one kilogram of liquid water 1 °C. It is equal to around 4.184 kilojoules. A larger calorie number indicates more energy. Large complex carbohydrates have higher calorie counts than small simple carbohydrates.

When complex carbohydrates are consumed, they are broken down by the body into simple sugars. These sugars pass through the intestinal wall and into the bloodstream. They are then carried throughout the body and delivered to cells. Once inside a cell, simple carbohydrates are broken down through cellular respiration into energy. If there are excess carbohydrates in the body, the unneeded carbohydrates are converted into fat molecules and stored in the body's tissues.

LIPIDS

Lipids are fats; they are made up of chains of methyl (-CH) units. The chains may be long or short. They may be straight or fused into rings (cyclic). They have several functions but are best known as molecules that store energy. They are also the structural components of the cell membrane. Go back to Chapter 5 and review the text that describes the structure of the molecules that make up the cell membrane. You will notice that in the cellular membrane, two fatty acid tails make up a large portion of that molecule. Figure 10.4 shows another lipid molecule — namely, a triglyceride.

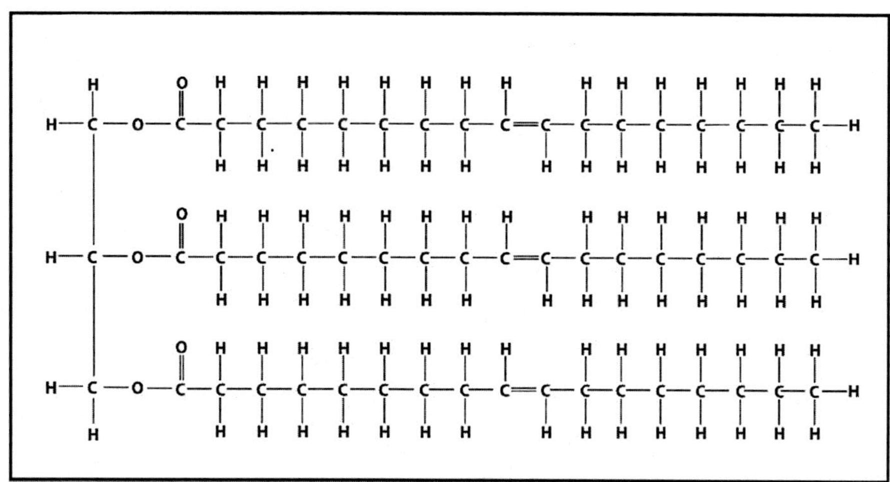

Figure 10.4 Lipid

After looking at Figure 10.4, can you guess why it is called a <u>tri</u>glyceride? That's right: it has **three** long carbon chains. Knowing that triglycerides are linked to heart disease, you may not be surprised to learn that butter contains triglycerides. Several important lipids have names that you may recognize: waxes, steroids, fatty acids and **triglycerides**. Lipid molecules consist of two main parts: a head and a fatty acid tail. The head is usually made up of carbons bound to hydrogens or oxygen. Recall that cellular membrane fats have a phospholipid structure. The head is made up of phosphate molecules bound to carbon molecules. Some fats, namely triglycerides, have a glycerol head. Glycerol is an organic molecule with alcohol (–OH) groups attached.

As we already mentioned, fats are a main part of the cell membrane. When eaten, fats are broken down through digestion and carried through the bloodstream to cells. Cells then use the individual glycerol or fatty acid molecules as energy or to create structural components of the cell. Fats are also used in the body as insulation. They surround and protect nerve tissue and major organs. Good fats should be a healthy part of your diet. Omega-3 and omega-6 fats are two examples of heart-healthy fats. Because fats usually have such long carbon-hydrogen chains, they typically contain the most calories per gram of any biomolecule.

NUCLEIC ACIDS

Nucleic acids are found in the nucleus of a cell. The nucleic acid polymer is made up of **nucleotide monomers**. The nucleotide monomer consists of a sugar, a phosphate group and a nitrogenous base. Nucleic acids are the backbone of the following genetic material:

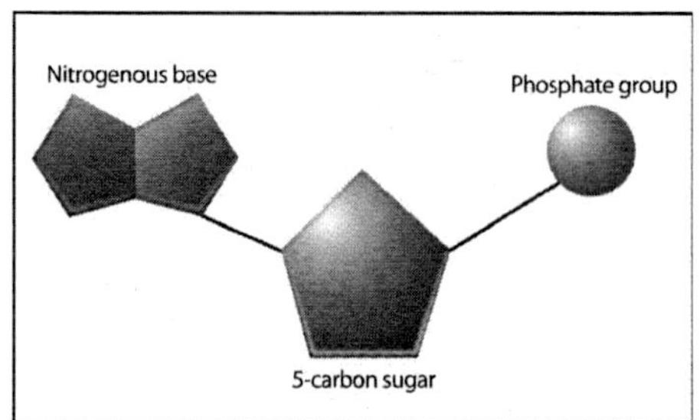

Figure 10.5 A Nucleotide

A **DNA** (deoxyribonucleic acid) directs the activities of the cell and contains the sugar deoxyribose.

B **RNA** (ribonucleic acid) is involved in protein synthesis and contains the sugar ribose.

PROTEINS

Proteins consist of long, linear chains of **polypeptides**. The polypeptide is itself a chain of **amino acid** monomers. There are 20 standard amino acids which combine to form every single protein needed by the human body. Figure 10.6 shows a polypeptide.

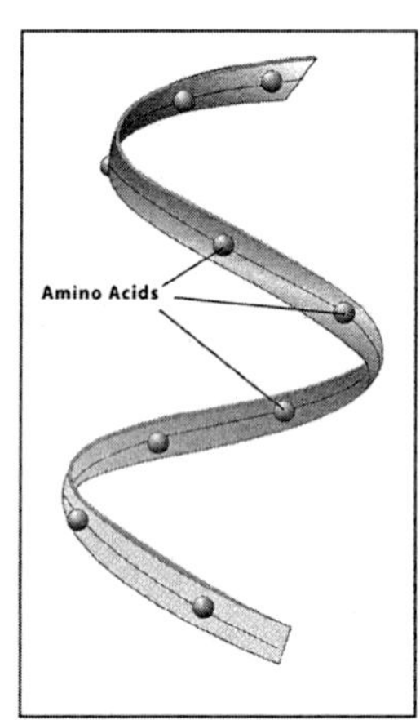

Figure 10.6 Polypeptide

There are many different types of proteins, all of which have different biological functions. They include: structural proteins, regulatory proteins, contractile proteins, transport proteins, storage proteins, protective proteins, membrane proteins and enzymes. Despite the wide variation in function, shape and size, all proteins are made from the same 20 amino acids. Recall the uniqueness of the side chain structures of amino acids discussed in Chapter 5.

Since mammals cannot make all 20 amino acids themselves, they must eat protein in order to maintain a healthy diet. Protein may be eaten in animal (meat) or vegetable (beans) form, but most organisms must have protein to survive.

During digestion, the long chains that make up proteins are broken down. Eventually, only the individual amino acid pieces remain. These monomers are absorbed into the bloodstream and are transported to various cells within the body. Inside a cell, the amino acid units are reassembled into various proteins needed by the organism. Inside bone marrow, amino acids are assembled into various blood proteins (one which you might recognize is hemoglobin). In another example, liver cells use amino acids to create various digestive enzymes. Glands use amino acids to create hormones. Hormones are protein messengers used within the body.

Section Review 1: Biomolecules

A. Define the following terms.

organic molecule	monomer	DNA	lipid	polymer
inorganic molecule	biomolecule	RNA	protein	polypeptide
	nucleic acid	carbohydrate	amino acid	

B. Select the best answer.

1 Carbon chains are principal features of **both** carbohydrates and lipids. What is the primary difference between these two types of biomolecules?

A Lipids always have longer carbon chains than carbohydrates.

B Carbohydrates carry hydroxyl groups on their carbon backbone.

C Carbohydrates cannot form rings as lipids can.

D Lipids provide energy, but carbohydrates do not.

2 Carbon is important to living things because

A it metabolizes easily, creating a quick energy source.

B it is abundant on the earth's surface.

C it can form four covalent bonds with other atoms.

D it has twelve protons and neutrons.

3 Nucleotides are to nucleic acids as amino acids are to

A DNA.

B lipids.

C proteins.

D carbohydrates.

C. Answer the following questions.

1 All living things have a common tie with the Earth on which we live. Explain why this is **true**.

2 What are the six elements commonly found in living things?

3 Why is carbon important to living things?

Catalysts and Enzymes

A **catalyst** is a substance that speeds up a chemical reaction without being chemically changed by the reaction. Catalysts decrease the amount of activation energy required for the reaction to occur. **Activation energy** is the amount of energy required in order for reactant molecules to begin a chemical reaction. When a molecule reaches its energy of activation, its chemical bonds are very weak and likely to break. Activation energy provides a barrier so that molecules will not spontaneously react with one another.

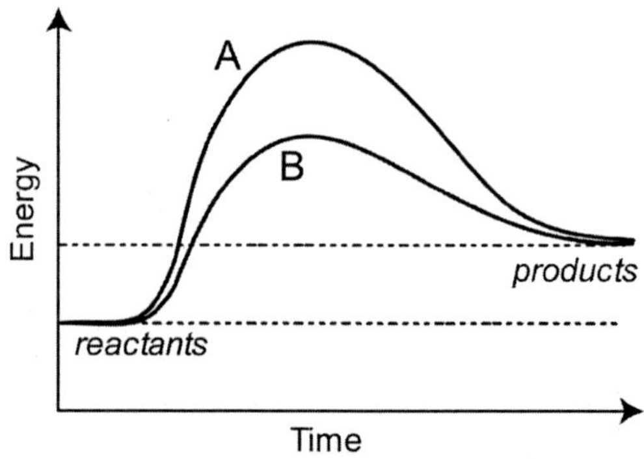

Figure 10.7 Effect of Catalysts on Activation Energy

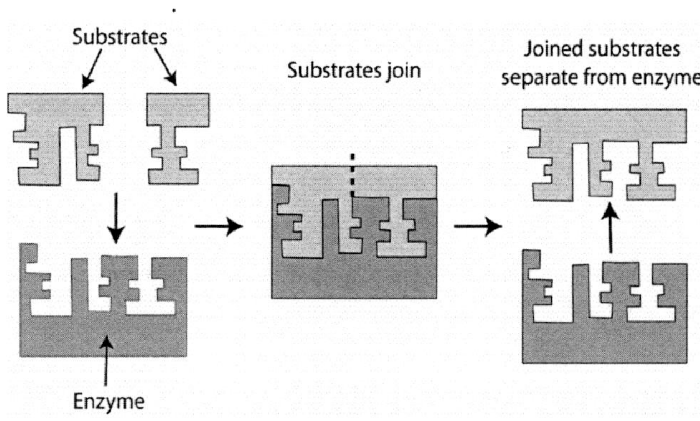

Figure 10.8 Enzymes

Our bodies use catalysts called enzymes to break down food and convert it to energy. Every cellular activity is a result of many biochemical reactions that take place at a cellular level. Substances that speed these reactions are called enzymes. **Enzymes** are specific proteins that combine with other substances called **substrates**. Substrates are the molecules acted upon by the enzyme. They are joined or broken depending on the chemical reaction taking place. There is one enzyme for one substrate, and they fit together like pieces of a puzzle. Metabolism cannot occur unless the energy of activation has been reached. These biological reactions would eventually take place on their own, but in the presence of enzymes, the reactions take place about a million times faster. Enzymes help to lower the energy of activation, making some chemical processes occur with greater frequency. Figure 10.8 shows a simplistic model of this reaction. You must understand that biological systems are much more complex.

Some reactions use **cofactors** to help enzymes by transporting electrons, ions or atoms between substances. A **cofactor** is an element or molecule that needs to be present, along with the enzyme in order for the reaction to take place. It may be a **metal ion** (a metal atom that has lost or gained electrons) or a coenzyme.

A **coenzyme** is a nonprotein molecule that activates the enzyme. Coenzymes can also be vitamins or made from vitamins. Important coenzymes are ATP (adenosine triphosphate) and **NAD+** (nicotinamide adenine dinucleotide). These coenzymes are used and reused during chemical reactions. ATP becomes ADP + NAD, which becomes NADP+. Coenzymes do not permanently bond to the enzyme but must be present for the enzyme to function. They often bring the energy needed to complete the reaction. We will not be addressing the specific movement of molecules and bonds in this text, but it is a good idea to have an idea of what these cofactors look like. Figure 10.9 shows the structure of the coenzyme NAD+.

Metabolic processes (like photosynthesis and respiration) can occur without enzymes, though at biological temperatures, metabolism would happen so slowly most organisms would be unable to survive. Some enzyme failures result in disease or death of the organism.

Figure 10.9 Coenzyme NAD+

Factors that influence the rate at which enzymes act include such things as temperature, pH, and amount of substrate present. Most enzymes have an optimum temperature and pH. Their optimum temperature or pH is the range at which the enzyme functions best. Enzymes vary from one organism to another. Some bacteria have enzymes that have an optimum temperature of 70 °C or higher; this temperature would destroy most human enzymes.

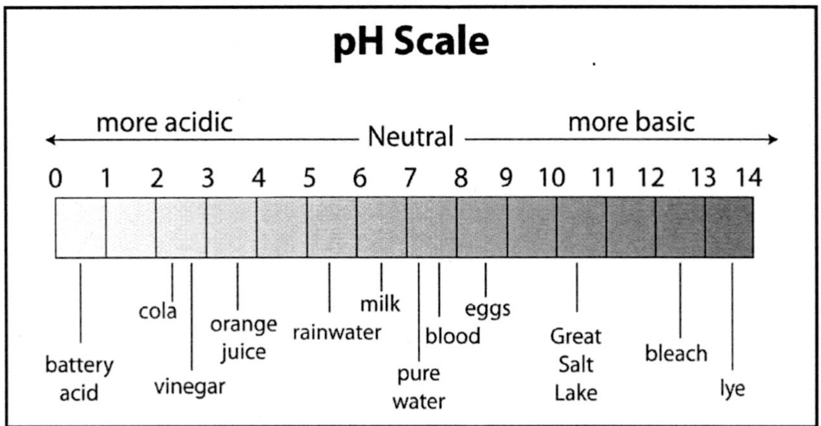

Figure 10.10 pH Scale

You will recall that a pH of 7 is considered **neutral**. Water has a pH of about 7. Substances with a pH less than 7 are **acids**, and substances with a pH greater than 7 are **bases**. With a few exceptions, most enzymes have an optimum pH of between 6 and 8. Figure 10.10 shows a pH scale. Recall from physical science class that pH stands for "potential of hydrogen." It is a measure of the H^+ or OH^- ions in a solution. Table 10.1 contains several enzymes and their optimal pH. One exception is pepsin, an acidic enzyme found in the human stomach. Pepsin has an optimum pH of 1 – 2. Outside their optimum pH, enzymes are broken apart, or denatured, by the acid or base.

Table 10.1 pH for Optimum Activity

Enzyme	Optimum pH
Lipase – hydrolyzes glycerides (pancreas)	8.0
Lipse – hydrolyzes glycerides (stomach)	4.0 – 5.0
Pepsin – decomposes proteins	1.5 – 1.6
Urease – hydrolyzes urea	7.0
Invertase – hydrolyzes sucrose	4.5
Maltase – hydrolyzes maltose to glucose	6.1 – 6.8
Amylase – hydrolyzes starch (pancreas)	6.7 – 7.0
Catalase – decomposes hydrogen peroxide into water and oxygen	7.0

Section Review 2: Cellular Energy

A. Define the following terms.

substrate	catalyst	enzyme	coenzyme
	activation energy	cofactor	pepsin

B. Choose the best answer.

1 What are enzymes?

A catalysts used by living things

B catalysts used in all reactions

C chemicals used to increase activation energy

D fats used by living things to help speed up chemical reactions

2 What is an example of a coenzyme?

A ATP B iron C chlorine D carbon

3 What is activation energy?

A The amount of energy required to start a chemical reaction.

B The amount of energy released by enzymes during a chemical reaction.

C The amount of energy found at the end of a chemical reaction.

D The amount of energy used to convert a cofactor into a coenzyme.

4 Which factor ***least*** affects enzyme function?

A temperature

B pH

C amount of substrate

D amount of UV light

5 Without enzymes, at which rate would metabolic reactions occur?

A very fast

B very slowly

C at a medium rate

D They would not occur at all.

C. Answer the following questions.

1 How are cofactors and coenzymes similar? How are they different?

2 In your own words, describe an enzyme.

3 Why is activation energy important?

Chapter 10 Review

1 Complex carbohydrates break down into which of the following?

A enzymes
B amino acids
C simple sugars
D ATP

2 Which of the following biomolecules are molecules that store energy with long chains?

A proteins
B nucleic acids
C carbohydrates
D lipids

3 Which of the following elements can be found in all living and previously living organisms?

A helium
B sulfur
C carbon
D nitrogen

4 Which biomolecule is a polymer assembled from some combination of the 20 amino acids?

A lipids
B DNA
C protein
D nucleotide

5 A coenzyme is a nonprotein molecule that activates the enzyme. What is the difference in the molecular structure of the protein and the coenzyme?

A A cofactor contains amino acids, but a protein does not.

B A protein contains amino acids, but a cofactor does not.

C A cofactor contains high-energy ionic bonds, but a protein does not.

D A protein contains high-energy ionic bonds, but a cofactor does not.

6 What are the largest carbohydrates called?

A monosaccharides
B disaccharides
C oligosaccharides
D polysaccharides

7 Lipids serve which function in a cell?

A store information
B store energy
C breakdown wastes
D join with substrates

8 What is the main goal of enzymes?

A lower the activation energy
B raise the activation energy
C slow the reaction rate
D find other molecules

9 What are proteins made of?

A long chains of hydrocarbons
B lipids
C amino acids
D salts

Biomolecules

10 What term **best** describes enzymes?

A accelerate

B decay

C retard

D dawdle

11 Which biomolecule determines the traits of organisms?

A proteins

B nucleic acids

C carbohydrates

D lipids

12 Which biomolecule must be consumed by animals?

A proteins

B nucleic acids

C carbohydrates

D lipids

13 The pancreas often secretes digestive enzymes to continue digestion. If a person eats a diet high in fat, what pH do you predict his pancreatic secretions will be?

A 4.5 – 5.0

B 7.0

C 6.7 – 7.0

D 8.0

14 What is a polymer?

A a short repeating subunit

B a long chain of repeating units

C an amino acid sequence

D an organic molecule

15 What is the ***most*** versatile biomolecule?

A protein

B nucleic acid

C carbohydrate

D lipid

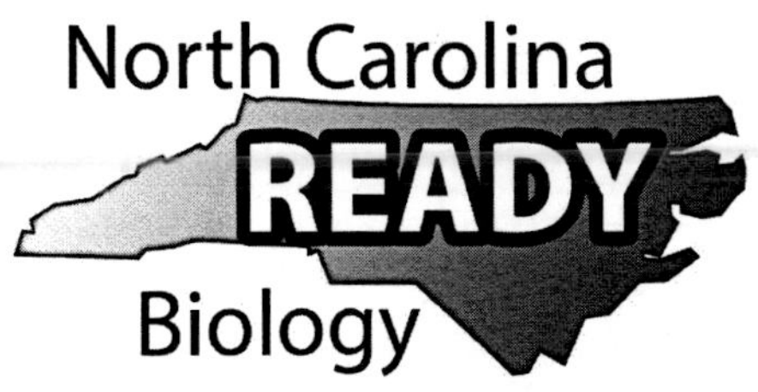

Chapter 11 Energy

This chapter covers:
4.2.1, 4.2.2

Energy Input for Ecosystems

Figure 11.1
Energy Supply for Ecosystems

The Sun has a strong influence on the Earth. It affects many things from weather to the northern lights. It even directs the everyday behavior of plants and animals. In fact, almost every living thing on Earth depends on the Sun for energy. Recall in chapter 3 that we discussed food chains and food pyramids. In these models, energy within ecosystems is passed through various levels from one organism to another. Any time energy is transformed, some is 'lost'. This is why energy within an ecosystem is often modeled using a pyramid shape. You might still be wondering — Where does the energy go? Well the answer is simple enough: living things must *live*. Meaning they must metabolize, grow and reproduce. All of these actions require energy. Animals have even higher energy requirements because they must move. Some energy is always being 'used' by organisms to live. This means that living actions almost always require more energy. For this reason, ecosystems require a continuous input of energy. If we think about this, it makes sense. Except in rare cases, plants, animals, bacteria and fungi all seek out a steady source of energy. For most organisms, the source of that energy is the Sun.

Keep in mind, the energy is not really lost, or destroyed, it only changes form. It is used to move cells, or molecules. It forms new tissues or catalyzes chemical reactions. The energy also becomes kinetic energy as animals move around the community.

In some ecosystems, plants trap energy from the Sun in a process called **photosynthesis**. Due to their energy-trapping ability, plants form the basis of most ecosystems. The radiant energy from the Sun predictably strikes the Earth's surface every day. Plants benefit from this predictability and use a unique energy-gathering technique called photosynthesis. The Sun's almost limitless energy source provides energy for almost every living creature on Earth. Plants use photosynthesis to trap energy, and animals eat plants for energy. In this chapter, we will learn more about the energy-capturing metabolisms of plants and animals.

ATP

ATP (adenosine triphosphate) is a molecule that serves as the chemical energy supply for all cells. Adenine, the sugar ribose and three phosphates compose ATP. The covalent bonds between the phosphate groups contain a great deal of energy. The release of that energy occurs when the last phosphate in ATP breaks off, forming **ADP (adenosine diphosphate)** and **P_i** (an inorganic phosphate molecule).

After the ATP molecule breaks down, ADP picks up free phosphate to form a new ATP molecule. Each ATP molecule is recycled in this way 2000 – 3000 times a day in the human body. The energy released during each cycle drives cellular processes. Examples of cellular processes that require energy include cellular transport, homeostasis, heat production, muscle contractions, photosynthesis, cellular respiration, locomotion, DNA replication and just about any other cellular process you can think of.

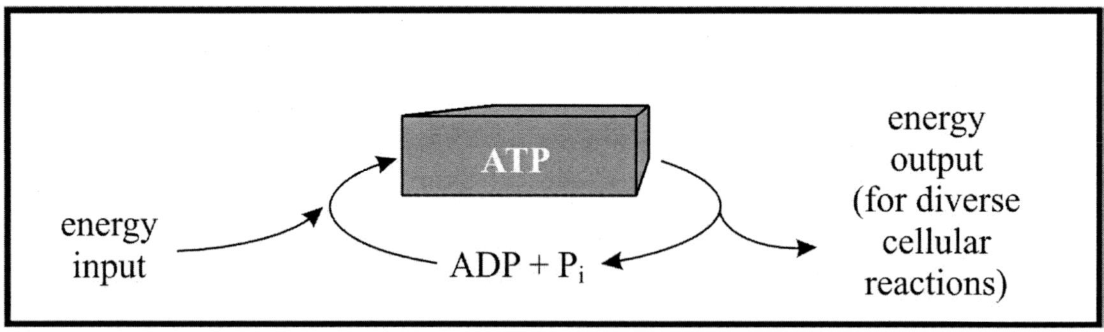

Figure 11.2 ATP/ADP Cycle

Obtaining Cellular Energy

Photosynthesis

Photosynthesis is the process of converting carbon dioxide, water and light energy into oxygen and high-energy sugar molecules. The chemical equation representing this process is shown in Equation 11.1. Plants, algae and some bacteria can use the sugar molecules produced during photosynthesis to make **complex carbohydrates** such as starch or cellulose for food. The process of photosynthesis consists of two basic stages: **light-dependent reactions** and **light-independent reactions**. The light-independent reactions are also called the **Calvin cycle**.

$$6CO_2 + 6H_2O + \text{light} \rightarrow C_6H_{12}O_6(\text{glucose}) + 6O_2 \qquad \textbf{Equation 11.1}$$

Photosynthesis takes place inside an organelle called a **chloroplast**. A chloroplast is one type of organelle called a plastid. **Plastids** engage in photosynthesis and store the resulting food. The chloroplast is a specific organelle with a double membrane that contains stacks of saclike membranes called **thylakoids**. The thylakoid membrane contains a green pigment called **chlorophyll**. **Pigments** are substances that absorb light. Light-dependent reactions take place inside the thylakoid membrane. Light-independent reactions take place in the **stroma** of the chloroplast. In the **light-dependent phase**, sunlight hits the leaf of the plant where it is absorbed by the pigments in the leaf. There are several plant pigments, but the most familiar is chlorophyll, the green pigment. Chlorophyll is stored in the chloroplasts of the plant cell.

When light hits the chlorophyll, electrons absorb the energy, become excited and leave the chlorophyll molecule. Carrier molecules transport the electrons, which follow an electron transport chain. Electron acceptor molecules pick up the electrons in a series and pass them from one molecule to another. As this occurs, energy is released, and ATP is formed. The final electron acceptor is NADP+.

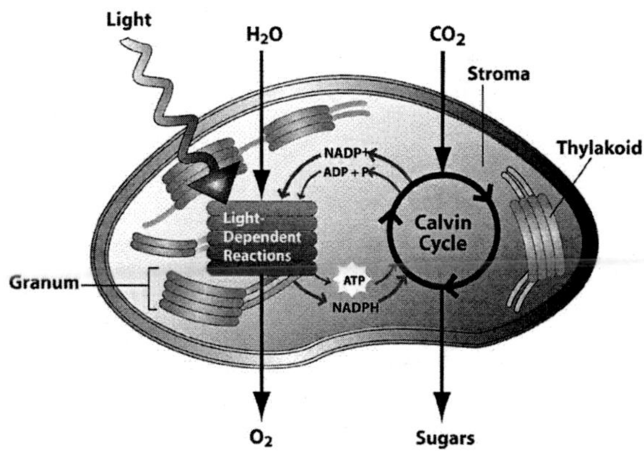

Figure 11.3 The Process of Photosynthesis

Splitting a molecule of water replaces the electrons released from the chlorophyll. These electrons, now available, combine with the NADP+ to form **NADPH**. The next stage of photosynthesis uses the NADPH, while oxygen is released as an end product of the reaction.

The end products of the light-dependent reactions are ATP, oxygen and NADPH. The ATP and NADPH are used in the light-independent reactions, and the oxygen is released into the atmosphere.

The next phase, the **light-independent** or **carbon fixation reactions**, uses the ATP formed during the light-dependent reaction as an energy source. In this phase, carbon (from carbon dioxide) and NADPH are used to form **glucose**. To accomplish this, a five-carbon sugar (called a pentose) uses a carbon atom from carbon dioxide to create a six-carbon sugar (a **hexose**). Glucose is the end result, after several conversions have taken place. The glucose can then be used as food to enter cellular respiration, or it can be converted to other carbohydrate products such as sucrose or starch.

Like any other chemical reaction, photosynthesis is also affected by external factors that can slow down or speed up the reaction. Let's first think about temperature. If it is very cold, then enzymes responsible for photosynthesis have very little kinetic energy. Meaning they move around very little. This, in turn, means they do not come into contact with the substrates needed to complete the reaction.

Additionally, at very low temperatures, water can freeze and become unavailable to complete the photosynthetic reaction. At very high temperatures, water again can evaporate quickly. Plants usually respond by closing their stoma (pores in leaves). This prevents rapid water loss, but it also stops the uptake of gases need for photosynthesis. Also at high temperatures, enzymes can break apart. As a result, if we were to graph the rate of photosynthesis compared to temperature, we would end up with the bell-shaped curve similar to Figure 11.4.

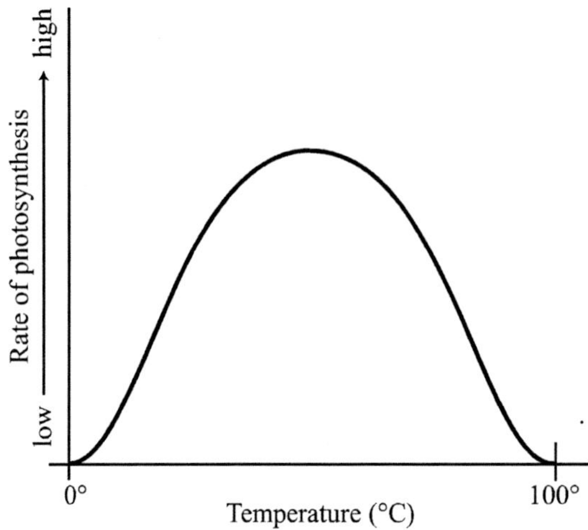

Figure 11.4 Rate of Photosynthesis vs. Temperature

In fact, other factors like pH, light level and amount of reactants also have an impact on the rate of photosynthesis. Like temperature, pH also shows a bell-shaped curve as an acidic solution or basic solution would break apart the photosynthetic enzymes. The amount of reactants and light levels have a positive relationship with the rate of reaction. The more reactants there are, the faster the reaction will progress. Figure 11.5 shows this relationship.

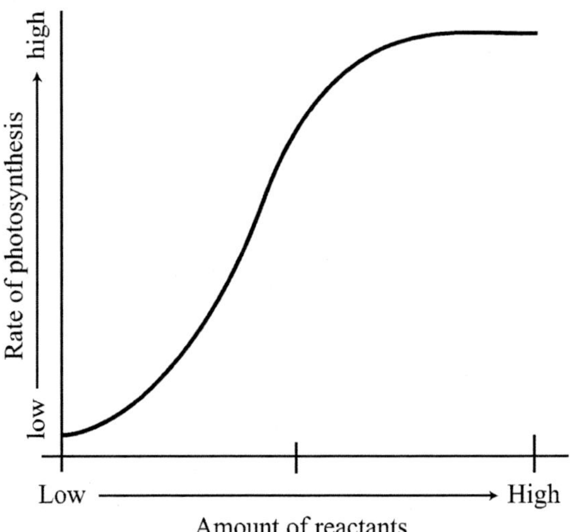

Figure 11.5 Rate of Photosynthesis vs. Reactants

Light-dependent photosynthetic reactions also respond positively to increased light levels.

Cellular Respiration

Cellular respiration is the process of breaking down food molecules to release energy. Plants, algae, animals and some bacteria use cellular respiration to break down food molecules. There are two basic types of cellular respiration: aerobic and anaerobic. **Aerobic respiration** occurs in the presence of oxygen, and is represented by the chemical equation in Equation 11.2. The energy released through cellular respiration is used to create ATP. Cellular respiration occurs in three phases: **glycolysis, Krebs cycle** and **electron transport**. The process starts with a molecule of glucose. The reactions of cellular respiration occur with the use of enzymes. Respiration is the primary means by which cells obtain usable energy.

$$C_6H_{12}O_6 + 6O_2 \rightarrow 6CO_2 + 6H_2O + \text{energy} \qquad \textbf{Equation 11.2}$$

Glycolysis is the first phase in cellular respiration. This step occurs in the cytoplasm of the cell, and it can occur whether or not oxygen is present. In this phase, the glucose molecule (a 6-carbon sugar) is broken in half through a series of reactions. The energy released by breaking down the glucose is used to produce ATP. Additionally, some high-energy electrons are removed from the sugar during glycolysis. These electrons pass on to an electron carrier called **NAD$^+$**, converting it to **NADH**. These electrons will later be used to create more energy.

In aerobic respiration, the 3-carbon sugars produced from glycolysis enter the **mitochondria** along with the oxygen. As the sugars enter the mitochondria, they convert to citric acid in phase two of cellular respiration. The **citric acid cycle**, or **Krebs cycle**, is the cyclical process that breaks down the citric acid through a series of reactions. The citric acid cycle produces more ATP. More high-energy electrons are released, forming NADH from NAD$^+$.

The last phase of cellular respiration is the **electron transport chain**, which occurs on the inner mitochondrial membrane. In this phase, the NADH releases the hydrogen ions and high-energy electrons it picked up during glycolysis and the citric acid cycle. The energy from these electrons is used to convert large quantities of ADP into ATP. The electrons transfer through a series of carrier proteins. At the end of the electron transport chain, the free electrons and H$^+$ ions bond with oxygen. The oxygen and H$^+$ ions form water, which is released from the cell as a waste product. Each electron transfer releases energy.

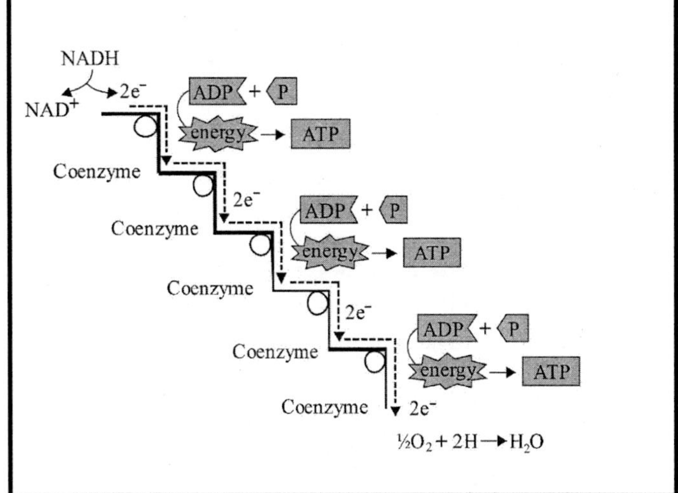

Figure 11.6 Electron Transport Chain

The multistep processes carried out during aerobic cellular respiration are very efficient. From the input of one glucose molecule, 36 ATP molecules can be produced.

Not surprisingly, respiration responds to changes in temperature and pH in much the same way as photosynthesis. Extremes in temperature or pH will slow the reaction as enzymes and substrates succumb to the severe environment. Again, we will notice a bell shaped curve when looking at a graph of respiration rate vs. pH.

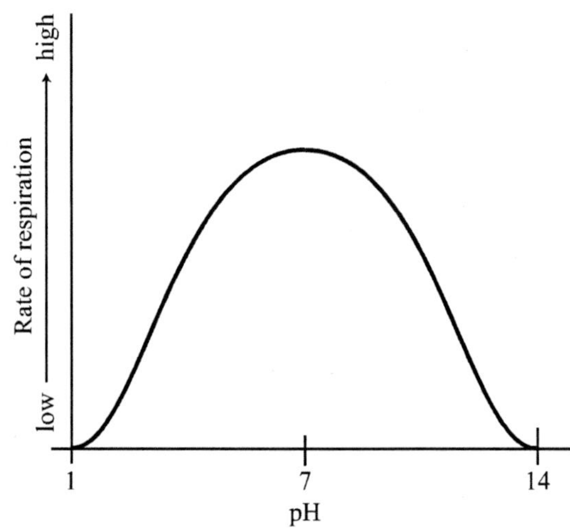

Figure 11.7 Cellular Respiration Rate vs. pH

Cellular respiration is relatively unaffected by light levels. Increasing the amount of reactants will increase the rate of the reaction.

Anaerobic respiration, or **fermentation**, is the process by which sugars break down in the absence of oxygen. Our muscle cells, fungi and some bacteria are capable of carrying out anaerobic respiration. These cells convert the products of glycolysis into either alcohol or **lactic acid.** Glycolysis releases energy, while the production of alcohol or lactic acid provides NAD^+, the electron carrier needed for glycolysis.

Yeast and some bacteria can carry out alcoholic fermentation. Yeast produces **ethanol** (C_2H_5OH) through **alcoholic fermentation**. The chemical equation representing this process is shown in Equation 11.3. During the fermentation process, carbon dioxide gas is released. This carbon dioxide gas is responsible for the holes in bread. The fermentation of yeast cells produces carbon dioxide, which becomes trapped in the dough, forming small bubbles and causing the bread to rise. Other uses of alcoholic fermentation are the making of beer, wine and liquor.

$$C_6H_{12}O_6 \rightarrow 2C_2H_5OH + 2CO_2 + \text{energy} \qquad \textbf{Equation 11.3}$$

Animal cells use a form of anaerobic cellular respiration called **lactic acid fermentation**. This generally occurs when the organism is engaging in strenuous exercise. During this period, the organism is not able to take in enough oxygen to meet the demand of all the cells in the body. So, in order to keep working, the cells begin to break down glucose in the absence of oxygen. The products of the reaction are energy and lactic acid.

Lactic acid fermentation proceeds at a very fast rate, but the amount of energy produced is far less than with aerobic respiration. Remember the 36 ATP molecules theoretically produced during aerobic respiration? By comparison, lactic acid fermentation produces only 2 ATP molecules.

So, can you tell if your cells are respiring aerobically or anaerobically? Well, you can assume that if you are doing anything that causes muscle fatigue — like weightlifting — those muscles need extra energy very quickly, and they obtain it through lactic acid fermentation. Your suspicions will be confirmed the next day if you are sore. Muscle soreness occurs because of lactic acid buildup in the muscle.

Activity

Use a sheet of paper to write by hand equations for photosynthesis and cellular respiration.

COMPARING PHOTOSYNTHESIS AND CELLULAR RESPIRATION

All organisms must be able to obtain and convert energy to carry out life functions, such as growth and reproduction. **Photosynthesis** is one way that organisms can trap energy from the environment and convert it into a biologically useful energy source. **Cellular respiration** is a way that organisms can break down energy sources to carry out life's processes. Photosynthesis takes place in plants, protists, algae and some bacteria. Cellular respiration takes place in all eukaryotic cells and some prokaryotic cells. It is important to mention that photosynthesis does NOT create energy, glucose or oxygen. It only *transforms* one type of energy into another type. Photosynthesis changes electromagnetic radiation energy from the Sun into chemical energy found in the bonds of the glucose molecule. This process very carefully changes one form of matter into another form. It takes the molecular carbon, hydrogen and oxygen and rearranges the matter into a different substance. Again, it does NOT create energy or matter — it only rearranges it.

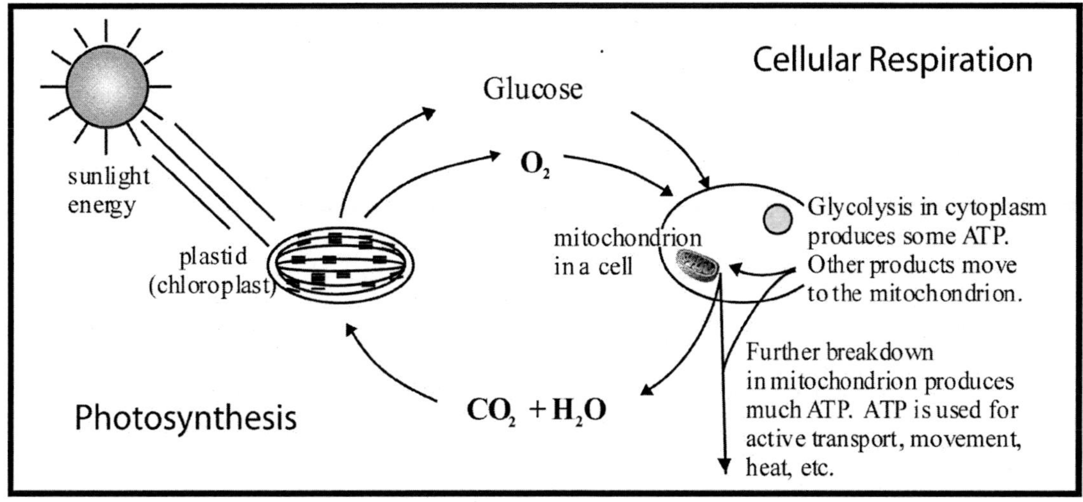

Figure 11.8 Relationship between Photosynthesis and Cellular Respiration

Table 11.1 Comparison of Photosynthesis and Cellular Respiration

	Photosynthesis	**Cellular Respiration**
Function	energy storage	energy release
Location	chloroplasts	mitochondria
Reactants	CO_2 and H_2O	$C_6H_{12}O_6$ and O_2
Products	$C_6H_{12}O_6$ and O_2	CO_2 and H_2O
Chemical Equation	$6CO_2 + 6H_2O + \text{light} \rightarrow C_6H_{12}O_6 + 6O_2$	$6O_2 + C_6H_{12}O_6 \rightarrow 6CO_2 + 6H_2O + \text{energy}$

Section Review 1: Obtaining Cellular Energy

A. Define the following terms.

photosynthesis
Calvin cycle
chloroplast
thylakoid
chlorophyll
pigment
plastid
carbon fixation
cellular respiration
aerobic respiration
anaerobic respiration
Krebs cycle
glycolysis
electron transport chain
ATP
ADP
P_i
mitochondria

B. Choose the best answer.

1 What form of energy is used by cells?

A enzymes B cofactors C ATP D DNA

2 The process of releasing energy from the chemical breakdown of compounds in a cell is

A hesitation. B expiration. C elimination. D respiration.

3 What is released when ATP is broken down into ADP and one phosphate?

A oxygen B water C energy D hydrogen

4 The process by which the energy from the Sun is used to create glucose molecules is known as

A cellular respiration.
B photosynthesis.
C chemosynthesis.
D fermentation.

5 How do plastids function within a cell?

A They digest food and break down wastes.

B They produce proteins.

C They carry on cellular respiration.

D They carry out photosynthesis and provide color.

C. Complete the following exercises.

1 Compare and contrast aerobic and anaerobic respiration.

2 Write the chemical equation for photosynthesis and cellular respiration using words instead of chemical formulas.

Active Transport

Think all the way back to chapter 2 when we discussed passive transport. You'll remember this is the method cells use to move molecules from an area of higher concentration to an area of lower concentration. As it turns out, cells have another way of transporting materials. This method moves materials from an area of lower concentration to an area of high concentration. To complete this process, the cell must expend energy. The movement of substances against the concentration gradient is called **active transport**. The movement is characterized by its directionality.

During active transport, cells can use special proteins, called carrier proteins, that are embedded within the cell membrane. Each carrier protein is specifically shaped to accommodate the type of molecule it transports. The molecule and the protein fit together like a lock and key. Carrier proteins use energy from ATP to physically change shape and move the molecule into the cell. Figure 11.9 shows the steps of active transport using carrier proteins.

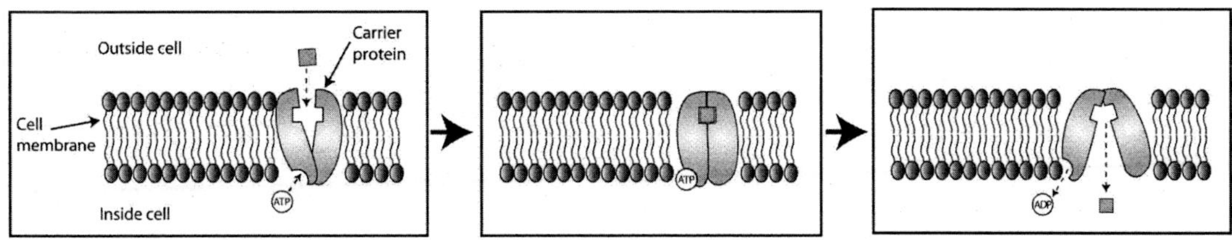

Figure 11.9 Active Transport

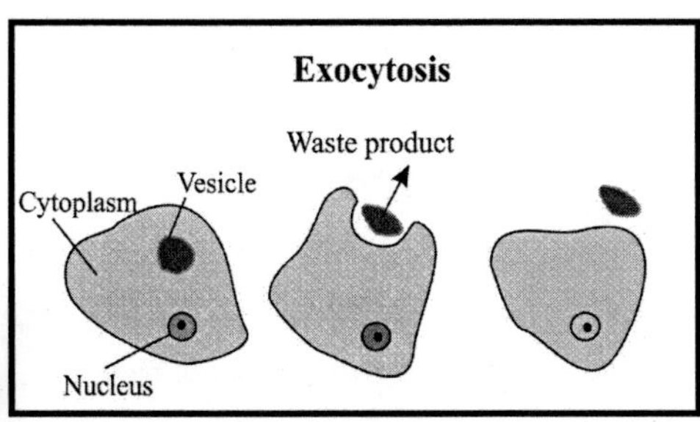

Figure 11.10 Exocytosis

Exocytosis is a form of active transport that removes materials from the cell. A sac stores the material to be removed from the cell, and then moves near the cell membrane. The cell membrane opens, and the substance is expelled from the cell. Waste materials, proteins and fats are examples of materials removed from the cell in this way.

Endocytosis, another form of active transport, brings materials into the cell without passing through the cell membrane. The membrane folds itself around the substance, creates a **vesicle**, and brings the substance into the cell. Some unicellular organisms, such as amoebas, obtain food this way.

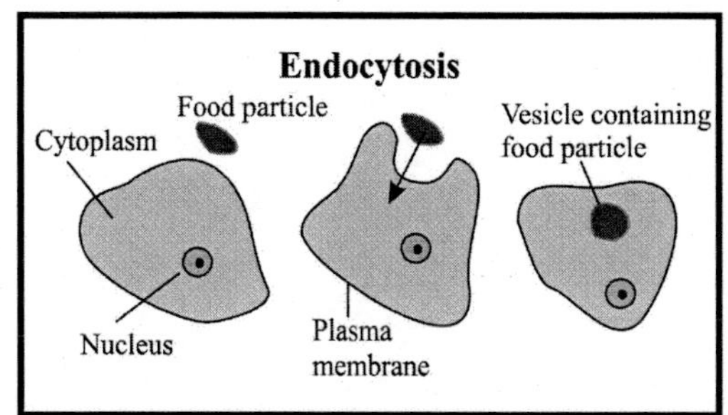

Figure 11.11 Endocytosis

Active transport is a mechanism that allows certain organisms to survive in their environments. For instance, sea gulls can drink salt water because their cells remove excess salt from their bodies through active transport. However, freshwater fish are not able to remove excess salt from their cells and, therefore, would become dehydrated in a saltwater environment. Another example of active transport involves blood cells, which use carrier proteins to transport molecules into the cell.

Section Review 2: Active Transport

A. Define the following terms.

hypotonic	active transport	endocytosis
homeostasis	exocytosis	vesicle

B. Choose the best answer.

1 The movement of substances into and out of a cell with the use of energy is called

A active transport.
C osmosis.
B passive transport.
D diffusion.

2 What is the movement of water across a semipermeable membrane from an area of high water concentration to an area of low water concentration called?

A active transport
C osmosis
B diffusion
D hypotonic

3 If the solution surrounding a cell has a higher concentration of solutes than inside the cell, what form of transport will the cell **most likely** use?

A active
B passive
C hypertonic
D hypotonic

C. Complete the following activities.

1 How does active transport differ from diffusion?

2 Dried beans are soaked overnight in preparation for cooking. Explain the process affecting the beans. What will happen to the dried beans?

3 Differentiate between exocytosis and endocytosis.

4 A celery stalk is placed in a solution. It begins to wilt. What is a **likely** component of that solution?

Chapter 11 Review

1 Amoebas obtain food by wrapping the cell membrane around the food particle, creating a vesicle. The food is then brought into the cell. What is this process called?

A exocytosis
B endocytosis
C osmosis
D photosynthesis

2 Which cellular component is instrumental in maintaining homeostasis?

A ribosome
B cytoplasm
C cell membrane
D cilia

3 In which kind of cell does cellular respiration take place?

A an animal cell only
B a plant cell only
C both plant and animal cells
D neither plant nor animal cells

4 To obtain and use cellular energy, plant cells use

A photosynthesis only.
B photosynthesis and cellular respiration.
C cellular respiration only.
D chemosynthesis.

5 What does the equation below represent?

$$C_6H_{12}O_6 + 6O_2 \rightarrow 6CO_2 + 6H_2O$$

A photosynthesis
B cellular respiration
C simple diffusion
D kinetic energy of motion

6 What is the **primary** purpose of photosynthesis?

A to trap and store solar energy in chemical form
B to release chemical energy from glucose into usable cellular energy
C to move molecules from one side of the membrane to the other
D to use chemical energy from water and oxygen

Energy

7 What biologically important molecule is a product of photosynthesis?

A O_2 B CO_2 C H_2O D $C_6H_{12}O_6$

8 Which term below **best** summarizes photosynthesis?

A capture

B free

C liberate

D disengage

9 Photosynthesis traps solar energy, which means it

A converts one energy form to another.

B makes energy in a unique form.

C breaks down energy.

D turns plants different colors in fall.

10 Which two organelles are involved in gathering and using energy within cells?

A mitochondria and vacuole

B vacuole and chloroplast

C chloroplast and mitochondria

D nucleus and ribosome

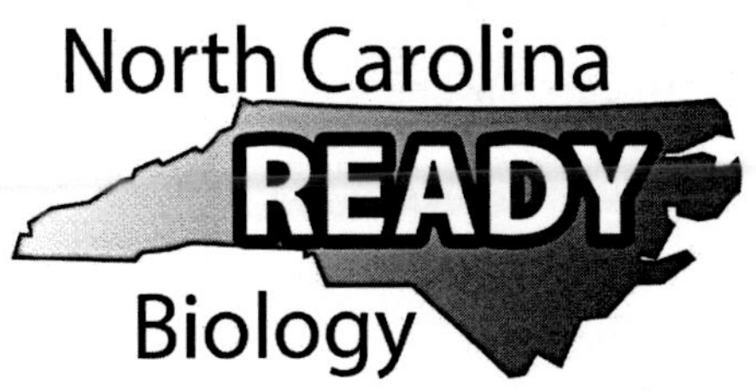

Post Test

1 A student made some observations about cells. She drew some diagrams to help her. 1.1.3

All of the cells she drew were from a multicellular organism *except* which cell?

A

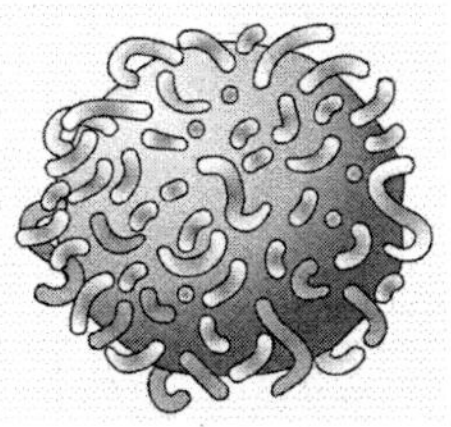

B

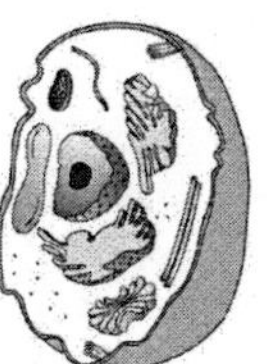

C

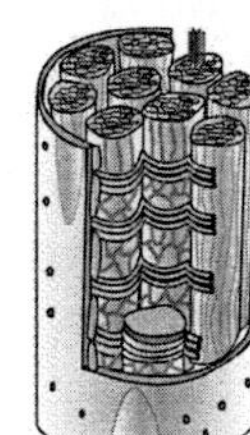

D

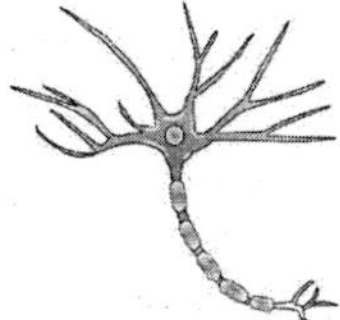

2 During seed germination, some cells display "light-seeking" traits while other cells develop "dark-loving" characteristics. Each type of cell takes on new physical characteristics that aid in the growth of the plant. What is this process known as? 1.1.3

A mitosis

B cell differentiation

C homeostasis

D cellular respiration

3 How are transpiration and evaporation related? 2.1.1

A Evaporation puts water into the water cycle, and transpiration removes water from the water cycle.

B Evaporation is the first step in the water cycle, and transpiration is the final step.

C They both represent ways in which water enters the atmosphere in the water cycle.

D They both represent ways in which water is removed from the atmosphere in the water cycle.

4 What is the ***main*** function of DNA replication? 3.1.1

A to copy genetic information

B to metabolize energy

C to store cellular energy

D to stop membrane transport

5 Scientists researched the lineage of a male cheetah named Spot. Spot's DNA sequence is: 3.3.1

AAT TAT CCG CTC

The DNA segments of possible offspring are shown in the table below. List the individuals in order from ***most*** related to ***least*** related.

Individual	DNA sequence
1	AAT TAT CCG CAG
2	TAG GAG ATC CAG
3	AAA TAC GGC CGG
4	TTA CAT CCG CTC

A 1, 4, 2, 3

B 1, 4, 3, 2

C 1, 2, 3, 4

D 2, 3, 4, 1

Go to next page

6 Which organism in the food chain has the ***greatest*** amount of available energy? 2.1.1

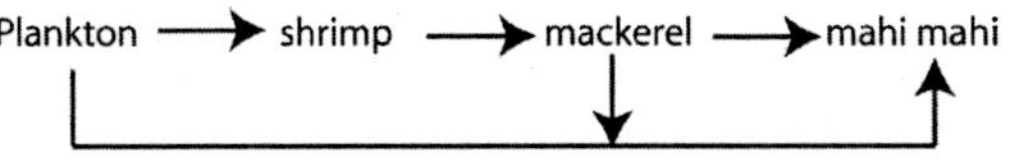

A shrimp

B plankton

C mackerel

D mahi mahi

7 A portion of mRNA has the sequence UUCAUGGGC. What was the sequence of the original DNA segment? 3.1.2

A AACTACCCG

B AAGUACCCG

C TTGTAGGGC

D AAGTACCCG

8 Recently, researchers have used a gene found in fish, called the fat-1 gene, to create a variety of omega-3 pigs. Using your knowledge of biotechnology, how do you think the researchers were able to produce this type of pig? 3.3.2

A selective breeding

B natural breeding

C gel electrophoresis

D recombinant technology

9 How do prokaryotic cells differ from eukaryotic cells? 1.1.2

A Prokaryotic cells are living, and eukaryotic cells are nonliving.

B Prokaryotic cells lack a true nucleus and membrane-bound organelles.

C Prokaryotic cells are much larger than eukaryotic cells.

D Prokaryotic cells require oxygen, and eukaryotic cells do not.

10 Trudy studied salamander behavior. She listed several common behaviors and the ages in which they were first seen in a table. 2.1.3

Behavior	Age First Seen
Swimming	Day 1
Limb Regeneration	Day 1
Lose Tail for Defense	Day 1
Avoid Stinging Insects	Day 25

Which behavior could ***best*** be described as a learned behavior?

A Swimming

B Limb Regeneration

C Lose Tail for Defense

D Avoid Stinging Insects

Go to next page

11 According to the diagram, which two organisms are the ***most*** related? 3.5.1

Organism	1	2	3	4
Common Name	Goatsbeard	Serviceberry	Prostrate Bluets	Purple Bluets
Phylum	Anthophyta	Anthophyta	Anthophyta	Anthophyta
Family	Rose	Rose	Madder	Madder
Genus	*Aruncus*	*Amelanchier*	*Houstonia*	*Houstonia*
Species	*dioicus*	*laevis*	*serpyllifolia*	*purpurea*

A Organisms 1 and 2

B Organisms 2 and 3

C Organisms 3 and 4

D Organisms 2 and 4

Go to next page

12 Which stage of mitosis is shown here? 1.2.2

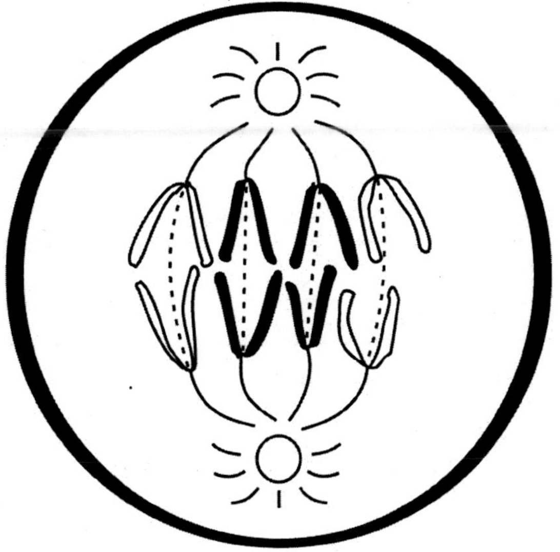

A prophase

B metaphase

C anaphase

D telophase

Go to next page

13 A scientists placed fish embryos in a magnesium chloride solution. Examine his data table below. 3.1.3

	Number of Eyes in Adult
Nutrient Solution	2
50 % Magnesium Chloride Solution	1
100 % Magnesium Chloride Solution	1

Which statement below ***best*** explains these results?

A Genes that code for eyes were destroyed.

B Genes that code for two eyes were mutated.

C Genes that differentiate light were switched off.

D Genes that code for facial traits were switched on.

14 Bacteria inhabit the intestines of cows. Cows eat plants but cannot digest the cellulose. The bacteria derive their nutrition from the plants the cows eat, and then they make available nutrients for the cow by breaking down the plant cellulose. 2.1.3

What symbiotic interaction is illustrated by the cows and bacteria?

A a relationship where both species are harmed

B a relationship where one species hunts and kills the other species

C a relationship where one species benefits and the other species is harmed

D a relationship where both species benefit and neither are harmed

Go to next page

15 A free-living unicellular organism reproduces asexually. A student made a diagram of its life cycle. 1.2.2

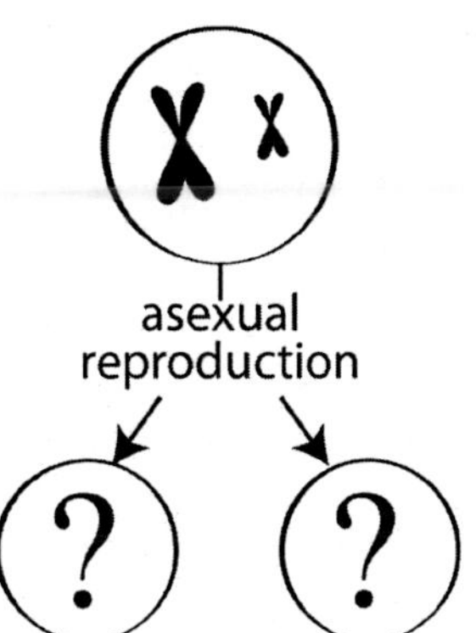

If the parent cell contains 2 chromosomes, how many chromosomes are contained within each daughter cell?

A 1

B 2

C 7

D 56

16 According to the classification key, to which genus and species does this member of the family Canidae belong? 3.3.1

1. a. Has a pointed muzzle.. Go to 2.
   b. Has a stout, blocky muzzle.. Go to 3.

2. a. Weighs 10 – 20 pounds .... *Vulpes vulpes*
   b. Weighs 20 – 50 pounds .... *Canis latrans*

3. a. Has yellow eyes and large teeth; undomesticated .... *Canis lupus*
   b. Has white eyes and smaller teeth; domesticated .... *Canis lupus familiaris*

A *Vulpes vulpes*

B *Canis latrans*

C *Canis lupus*

D *Canis lupus familiaris*

Go to next page

17 Normally frightened by loud noises, fish living near a marina are attracted to the sounds of boat motors. Returning boats often dump food scraps into the water. What type of learning is demonstrated by the fish? 3.4.2

A imprinting

B trial and error

C migration

D habituation

18 Which biomolecules, made by cells, are the ***most*** versatile and responsible for many cell functions? 4.1.1

A fats

B carbohydrates

C proteins

D catalysts

19 Which molecule carries information from the DNA in the nucleus out into the cytoplasm of the cell? 4.1.2

A tRNA

B rRNA

C ATP

D mRNA

20 In aspen trees, the allele for having round leaves (R) is dominant to the allele for having oval leaves (r). Use the Punnett square to determine the probability of heterozygous parent trees having offspring with round leaves. 3.2.2

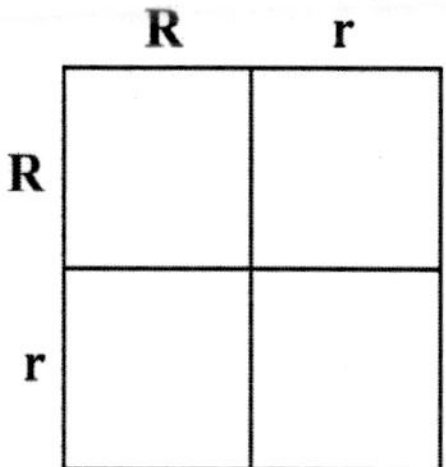

A 75%

B 25%

C 50%

D 0%

21 Two parents have four children. Three of the children have a widow's peak and one child has no widow's peak. 3.2.2

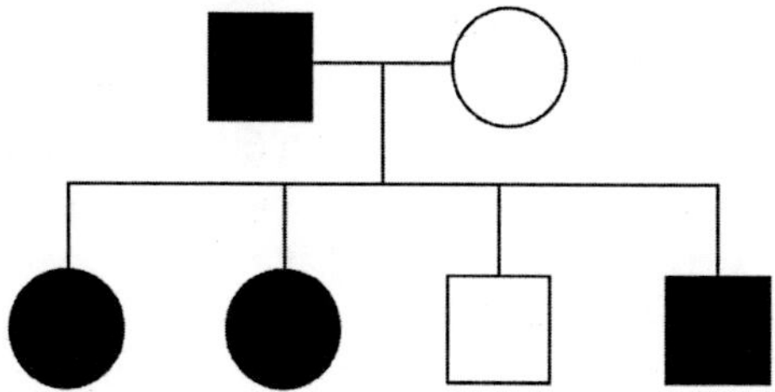

What is the genotype of the parent who has a widow's peak?

A WW

B Ww

C ww

D Wp

Go to next page

22 Research students tagged the amino acid base uracil with a radioactive tag. They then measured the amount present in three samples. 3.1.2

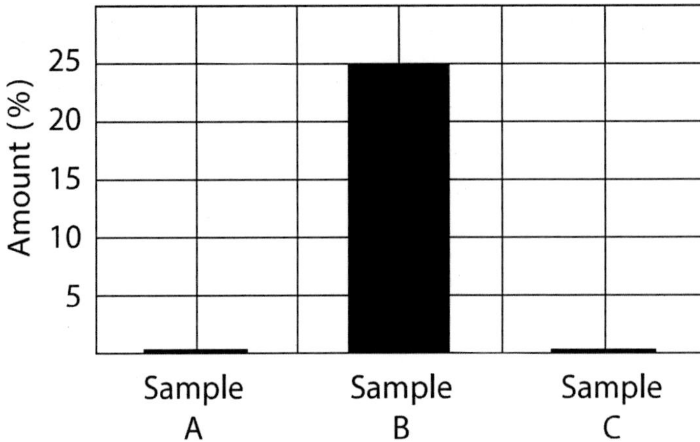

Which statement below is a valid conclusion?

A Sample A was an RNA molecule.

B Sample B was an RNA molecule.

C Sample B was a DNA molecule.

D Sample C was an RNA molecule.

23 After fertilization, an embryo develops into a zygote through many cell divisions. If the sperm and egg each contain 8 chromosomes, how many chromosomes are contained within the zygote? 3.2.1

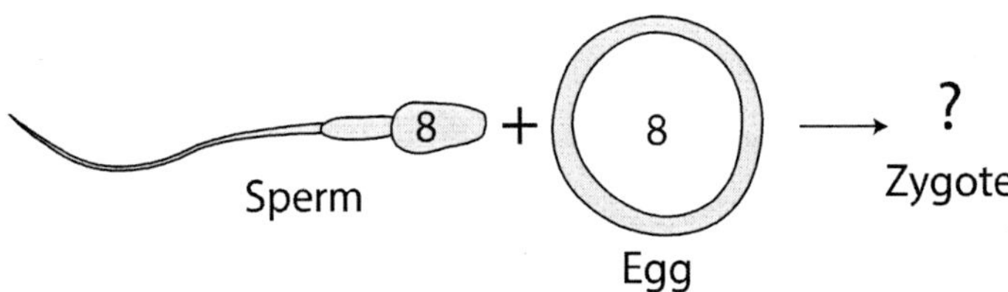

A 4

B 8

C 16

D 32

Go to next page

24 What is the inheritance pattern for color in pansies? 3.2.2

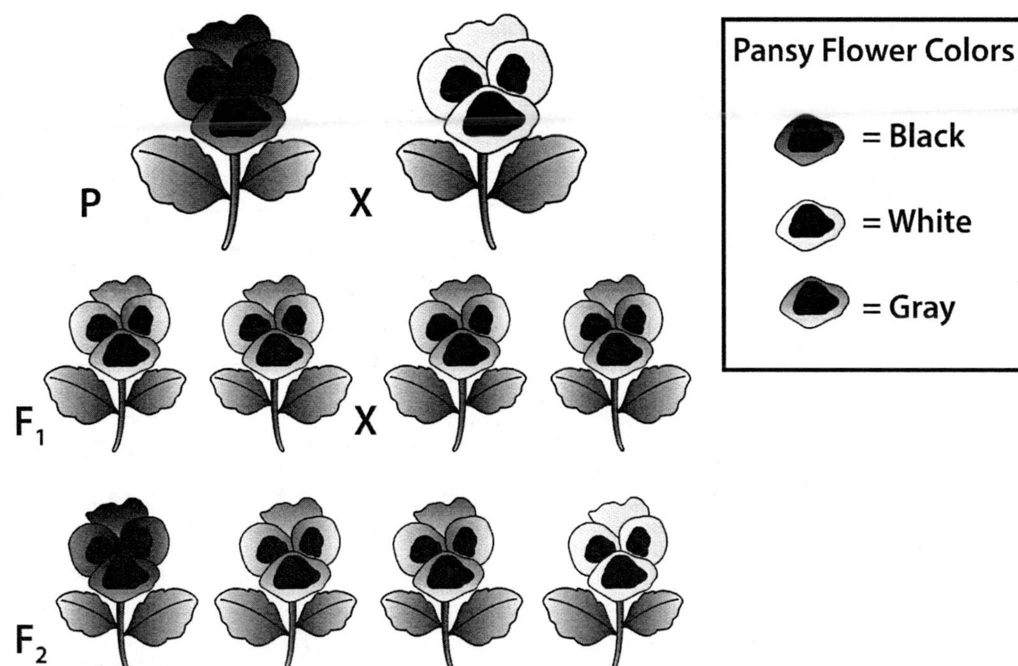

A simple dominant/recessive

B sex-linked

C codominant

D incomplete dominant

25 Mistletoe extracts water and nutrients from several types of trees, harming the tree. What type of symbiotic relationship is illustrated by the action of mistletoe? 2.1.3

A mutualism

B predation

C parasitism

D commensalism

26 What is one benefit of genetically engineered plants? 3.3.2

A They are renewable resources.

B They allow crops to be grown in many climates.

C They use chemicals that harm the Earth.

D They cause allergies.

27 A cell is capable of carrying out both photosynthesis and respiration. How would it ***best*** be described? 4.2.1

A prokaryotic cell

B mitotic cell

C plant cell

D animal cell

28 What is the term for a molecule that speeds up a chemical reaction but is ***not*** changed by the reaction? 4.1.3

A product

B reactant

C enzyme

D protein

29 The spines of a cactus are modified leaves. The thorn of a rose is a modified branch. What does this suggest about the evolution of these two families of plants? 3.4.1

A The spine and the thorn are homologous structures and are proof of common ancestry.

B The spine and the thorn are analogous structures and are not proof of common ancestry.

C The spine and the thorn have separate functions, so they are not homologous and provide no evidence to support a common ancestor.

D The spine and the thorn are vestigial structures that have not evolved.

Go to next page

30 Fabio did an experiment involving the growth of plants. Seedlings were planted in pots and then set up as shown in the diagram below. The diagram shows the growth of the plants as of week 16 of the experiment. All of the plants were planted in the same type of soil, given the same amount of light and water, and were kept at a constant temperature. Based on the diagram below, which one of the following statements is a valid conclusion that Fabio can draw from his experiment? 2.1.2

A The plants used in the experiment tend to grow toward their light source.

B The plants used in the experiment will always tend to grow against the force of gravity.

C All plants need a light source.

D The size of the plant is a function of the amount of light it receives.

31 How do pesticides function within the context of natural selection? 3.4.2

A Pests become immune to chemical controls and survive.

B Despite interference from pesticides, humans survive.

C Antibiotics destroy microbes in the environment.

D Chemical compounds within plants increase in potency to destroy pests.

32 What does the DNA code direct the cell to manufacture? 4.1.2

A protein

B amino acids

C hydrogen bonds

D sugars

33 Sewage treatment facilities sometimes allow tons of untreated sewage to flow into river systems. Which of the following is an appropriate method for dealing with this situation? 2.2.1

A build dams along the river to catch the sewage

B put up signs warning of pollution

C process the untreated sewage by adding oxygen and sewage-eating bacteria

D ask farmers in the area to use some untreated sewage for fertilizer

34 The family Elephantidae contains both living and extinct species of the largest land mammal on Earth. Examine the pictures below comparing their ear sizes. 3.4.2

Mummified Mammoth

African Elephant

What is the ***most likely*** explanation for the variations in ear length in these different species?

A differences in diet led to smaller sized individuals

B differences in climate led to distinctive body shapes

C the process of fossilization caused the appendages of the mammoth to shrink

D living tissues are full of water thus swell to a larger size

35 A raisin was left in a solution overnight and was swollen by morning. The solution that the raisin was placed in was 3.5.2

A hypertonic.

B hypotonic.

C isotonic.

D acidic.

Go to next page

36 Examine the lifespan diagram to answer the question. 3.4.2

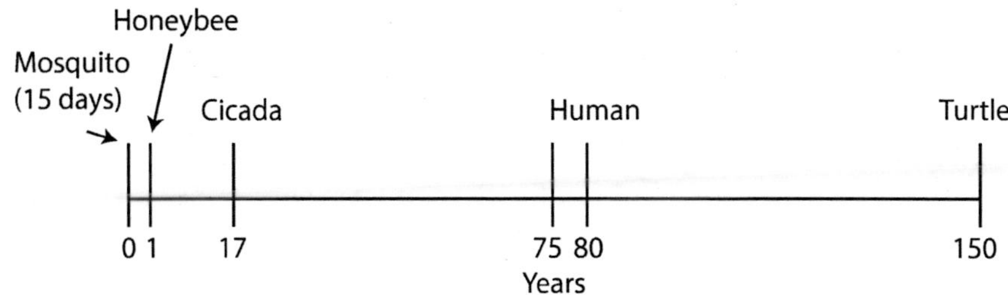

Of these organisms, which do you predict would have the highest rate of evolution?

A mosquito

B honey bee

C turtle

D cicada

Go to next page

37 A species of bird lives in a canyon. This bird is reproductively isolated from other bird species in the area. The males of this type of bird species produce colorful red and blue feathers. During warmer years (greater than 78 °F), more food is available, and males can produce more vivid coloration. Females will only mate with males that have vivid coloration. During these warmer years, there are more successful breeding pairs of birds and more offspring. The graph below summarizes the number of offspring born each year. 2.1.2

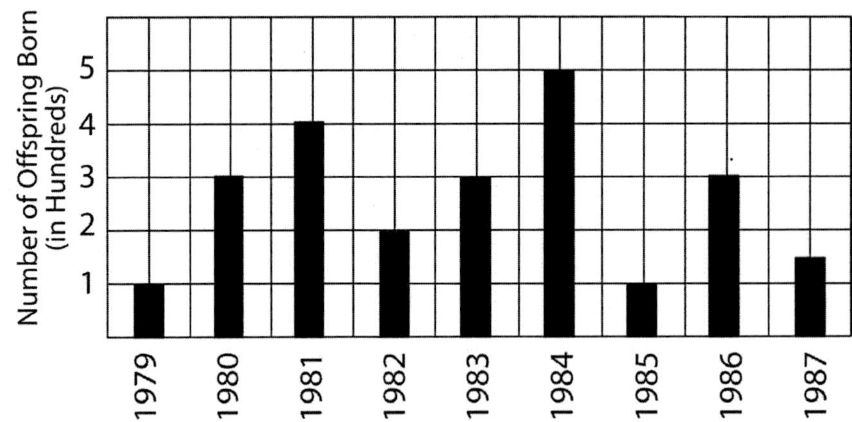

If the average breeding season temperature is greater than 80 °F for the next five years, what trend will be observed?

A Bird coloration will become more dull.

B Bird coloration will stay the same.

C More offspring will be produced.

D Only female birds will be produced.

38 Scientists classify the raccoon and the rabbit as the most related animals in this table. What two traits ***best*** support this classification? 3.5.1

Characteristics of Four Animals

Organism	Native Range	Foot Shape	Body Covering	Active	Diet
Raccoon	N. America	Nonwebbed	Fur	Night	Omnivore
Duck	N. America	Webbed	Feathers	Day	Herbivore
Dyeing Dart frog	S. America	Webbed	Skin/Water	Day	Carnivore
Rabbit	N. America	Nonwebbed	Fur	Night	Herbivore

A foot shape and body covering

B native range and activity

C body covering and diet

D foot shape and diet

Go to next page

39 The equation below summarizes which biological process? 4.2.1

Light energy + $6H_2O + 6CO_2 \rightarrow C_6H_{12}O_6 + 6O_2$ +ATP

A chemophotosynthesis

B fermentation

C photosynthesis

D cellular respiration

40 The snowshoe hare grows a white winter coat. This chromatic camouflage hides it from the fox, an important predator. During the summer, the hare grows a brown coat. If unusually warm winter conditions cause premature melting of the snow, what would you expect to happen to the rabbit population? 2.1.3

A It would increase greatly due to the increased food supply.

B It would decrease greatly due to the increased predation.

C It would probably increase slightly, with greater food supplies for both the rabbit and the fox being the deciding factor.

D It would probably decrease somewhat, with increased predation outweighing the effect of greater food supply.

41 Which process listed below produces the ***most*** ATP molecules? 4.2.1

A lactic acid fermentation

B alcoholic fermentation

C aerobic respiration

D protein synthesis

42 Animal cells that are specialized to transmit messages are ***most likely*** found in what part of the animal? 1.1.3

A the nervous system

B the circulatory system

C the skeletal system

D the muscular system

43 What is the ultimate goal of the process shown in the diagram below? 3.1.2

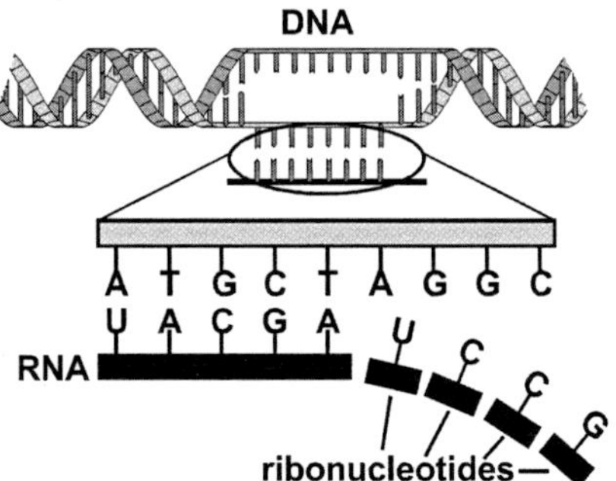

A to store cellular energy

B to maintain homeostasis

C to replicate DNA

D to make protein

Go to next page

44 The organelle indicated by an arrow in the diagram captures sunlight. 4.2.1

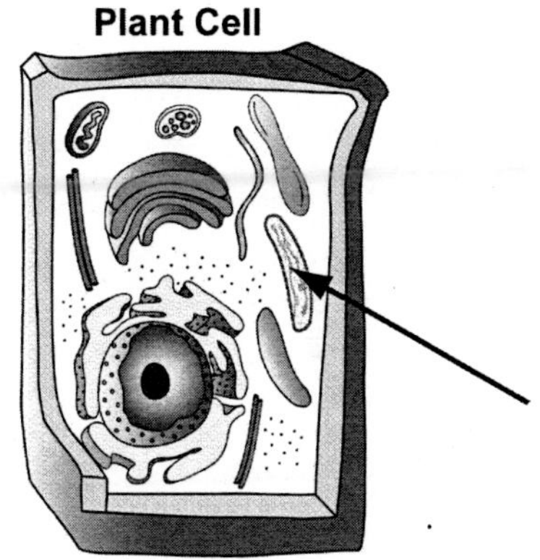

What process does the organelle complete for the plant cell?

A photosynthesis

B aerobic respiration

C nutrient absorption

D cellular transport

Go to next page

45 A cellular process is shown in the diagram below. 3.2.1

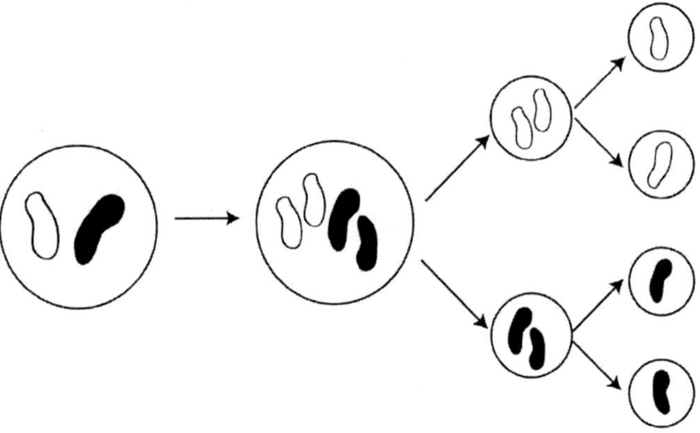

What does this process ***most likely*** increase in populations?

A genetic variation

B genetic mutation

C fertilization rates

D the rate of evolution

46 A plant has a thick, waxy cuticle to prevent moisture loss. The interior of the plant is hollow and is used to store large quantities of water. The leaves of the plant have evolved into sharp spines, which protect the flesh of the plant from water-seeking animals. Which environment is ***most*** suited to this organism? 2.1.2

47 Which type of bird foot is ***best*** adapted to walking on the ground? 2.1.2

A

C

B

D

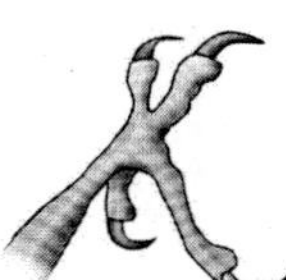

Go to next page

48 A scientist treats a cell with a chemical that helps lysosomes. As a result, which cell process will be ***most*** helped? 1.1.1

A protein synthesis

B active transport

C waste removal

D mitosis

49 Which controversial biotechnology offers the best hope to end global malnutrition? 3.3.3

A cloning

B gene therapy

C Human Genome Project

D genetically modified organisms

50 Which habitat is ***best*** suited for an amoeba? 1.2.3

A a glacier

B a desert

C a swamp

D a swift river

51 Which organism listed below is ***most likely*** able to carry out anaerobic respiration? 4.2.1

A a portabella mushroom

B a maple tree

C an azalea bush

D a tomato plant

52 Vscan™, developed by General Electric, is a new pocket-sized ultrasound device. This device, no larger than a flip phone, will allow doctors to see a patient's internal organs at a glance. Which step will ***most*** help the company determine if the pocket-sized ultrasound machines are beneficial to doctors and patients? 3.3.1

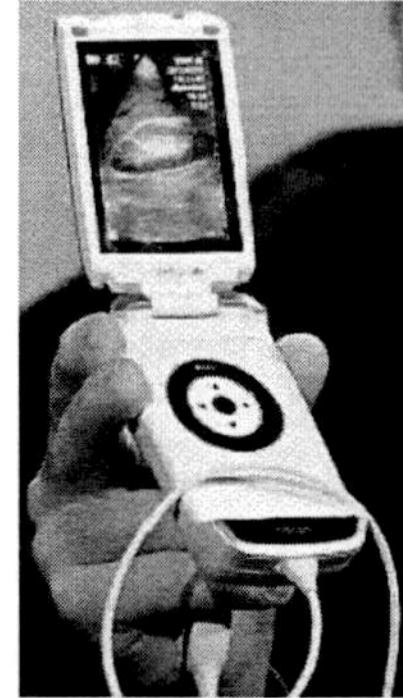

A developing color cases for the device

B upgrading the device for use with wireless technology

C testing the device in various hospitals around the world

D telling children about the device

Go to next page

53 Which image below ***best*** represents a eukaryotic cell? 1.1.2

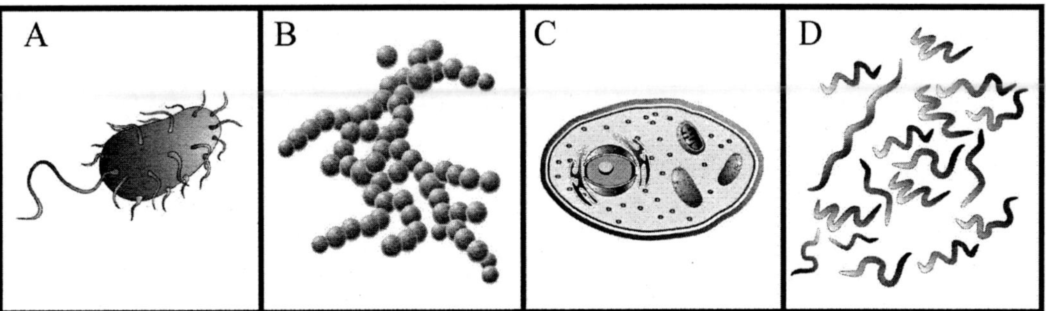

54 The removal efforts of the Water Chestnut Eradication project attempted to remove the invasive plant from American river systems. In 1999, approximately 200,000 lbs. of water chestnuts were removed from Maryland waterways. In 2000, around 1,000 lbs. were removed. In 2001, less than 500 lbs. were removed. Which graph below correctly represents this data? 2.2.2

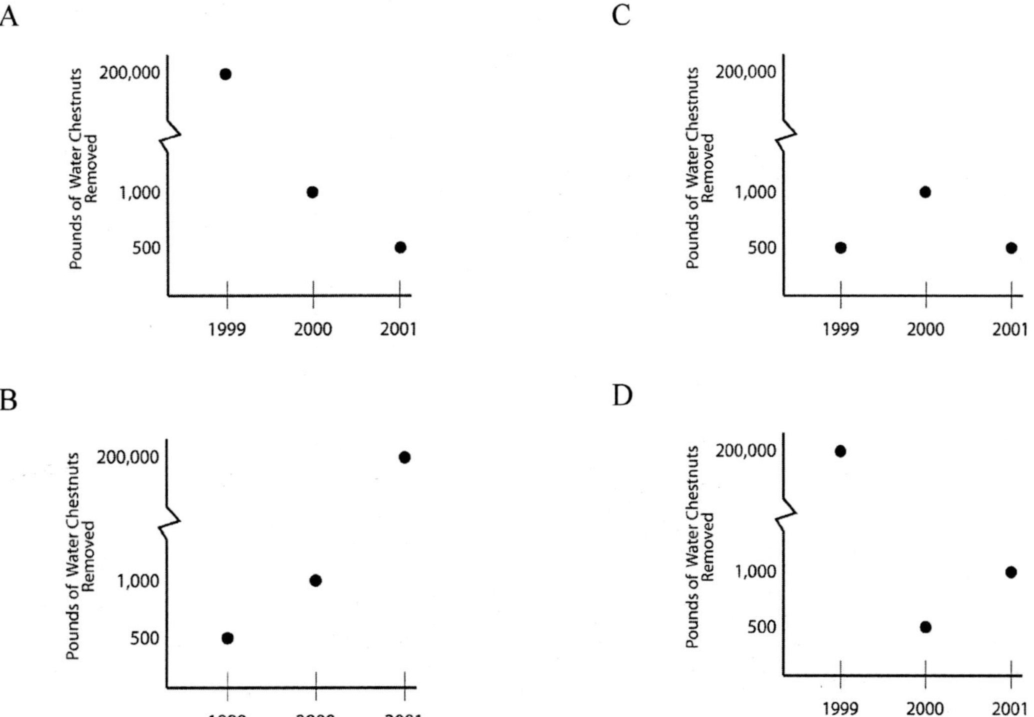

Go to next page

55 Plants produce oxygen, which animals use. Animals produce carbon dioxide, which plants use. Predict the ***most likely*** outcome if airborne pollution blocked the Sun's rays so significantly that the percentage of living plants was decreased by 95%. 1.2.2

A Plants would quickly evolve to consume a mixture of oxygen and carbon dioxide.

B Animals would quickly evolve to begin using carbon dioxide.

C The atmosphere would fill with oxygen, and animals would begin to die.

D The atmosphere would fill with carbon dioxide, and animals would begin to die.

56 Which pair below ***best*** represents a size model comparing a prokaryotic cell to a eukaryotic cell? 1.1.2

A a basketball and a soccer ball

B a baseball and a tennis ball

C a marble and a gum ball

D a golf ball and a volleyball

57 The DNA strand below is one half of a complementary pair. 3.1.1

TACCCATTCGAT

Which correctly shows the complementary pair to the DNA strands?

A TACCCATTCGAT

B TACCCATTCTAG

C UACCCAUUCGA

D ATGGGTAAGCTA

Go to next page

58 Tandra discovered that temperature inside a reptile nest determines the sex of the offspring. She placed eggs from a single box turtle species in an incubator at various temperatures. Examine her data below. 3.2.3

	Temperature (°F)	Resulting Sex
Clutch A	85	100 % Male
Clutch B	87	50 % Male; 50 % Female
Clutch C	90	100 % Female
Clutch D	95	No Eggs Hatched

What is the ***best*** conclusion based on this data set?

A The nest environment has little to no influence on the sex of box turtles.

B The nest environment is most important in sex determination of box turtles.

C Nests kept at 85 °F or cooler will create only female box turtles.

D Nests kept at 90 °F or warmer will create only male box turtles.

59 Examine the diagram to answer the question. 1.1.3

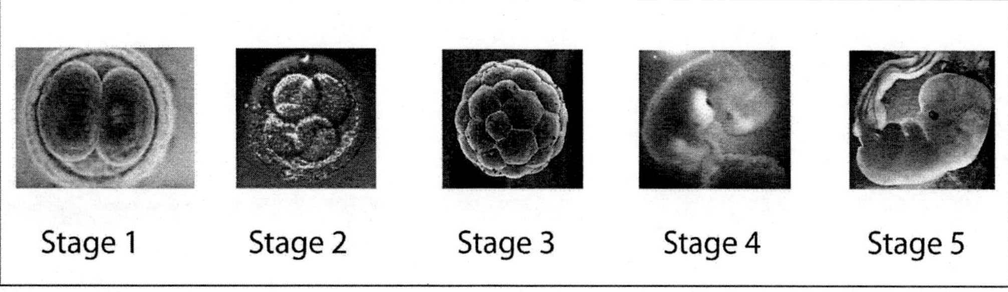

Cell differentiation occurred between which two stages?

A Stage 1 and 2

B Stage 2 and 3

C Stage 3 and 4

D Stage 4 and 5

60 This chart shows a list of messenger RNA codons. 3.1.2

Second Position

First Position	U	C	A	G	Third Position
U	Phenylalanine	Serine	Tyrosine	Cysteine	U
	Phenylalanine	Serine	Tyrosine	Cysteine	C
	Leucine	Serine	Stop	Stop	A
	Leucine	Serine	Stop	Tryptophan	G
C	Leucine	Proline	Histidine	Arginine	U
	Leucine	Proline	Histidine	Arginine	C
	Leucine	Proline	Glutamine	Arginine	A
	Leucine	Proline	Glutamine	Arginine	G
A	Isoleucine	Threonine	Asparagine	Serine	U
	Isoleucine	Threonine	Asparagine	Serine	C
	Isoleucine	Threonine	Lysine	Arginine	A
	Methionine	Threonine	Lysine	Arginine	G
G	Valine	Alanine	Aspartic acid	Glycine	U
	Valine	Alanine	Aspartic acid	Glycine	C
	Valine	Alanine	Glutamic acid	Glycine	A
	Valine	Alanine	Glutamic acid	Glycine	G

A strand of DNA with the sequence CCA GTA AAC will code for which amino acid sequence below?

A proline-valine-asparagine

B glycine-histidine-leucine

C glutamic acid-arginine-cysteine

D histidine-glutamine-phenylalanine